D0905386

Robust Statistical Procedures

PETER J. HUBER
Eidgenössische Technische
Hochschule Zürich

SOCIETY for INDUSTRIAL and APPLIED MATHEMATICS

PHILADELPHIA, PENNSYLVANIA 19103

First printing 1977.
Second printing 1979.

ISBN: 0-89871-024-3

Printed for the Society for Industrial and Applied Mathematics by
J. W. Arrowsmith, Ltd., Bristol 3, England

Contents

Chapter VII
ADAPTIVE ESTIMATES

Preface

At the NSF/CBMS Regional Conference at Iowa City, 19–23 July 1976, I gave ten consecutive talks on robust statistical procedures. These lecture notes follow very closely the actual presentation; their preparation was greatly facilitated by the excellent notes taken by J. S. Street during the conference. There was neither time nor space to polish the text and fill out gaps; I hope to be able to do that elsewhere.

All participants of the conference will remember Bob Hogg's impeccable organization and the congenial atmosphere. Many thanks also go to Carla Blum who did overtime typing the manuscript.

PETER J. HUBER

Zurich, May 1977

CHAPTER I

Background

1. Why robust procedures? The word "robust" is loaded with many—sometimes inconsistent—connotations. We shall use it in a relatively narrow sense: for our purposes, *"robustness" signifies insensitivity against small deviations from the assumptions.*

Primarily, we shall be concerned with distributional robustness: the shape of the true underlying distribution deviates slightly from the assumed model (usually the Gaussian law). This is both the most important case (some of the classical statistical procedures show a dramatic lack of distributional robustness), and the best understood one. Much less is known what happens if the other standard assumptions of statistics are not quite satisfied (e.g. independence, identical distributions, randomness, accuracy of the prior in Bayesian analysis, etc.) and about the appropriate safe-guards in these other cases.

The traditional approach to theoretical statistics was and is to optimize at an idealized model and then to rely on a continuity principle: what is optimal at the model should be almost optimal nearby. Unfortunately, this reliance on continuity is unfounded: the classical optimized procedures tend to be discontinuous in the statistically meaningful topologies.

An eye-opening example has been given by Tukey (1960):

Example. Assume that you have a large, randomly mixed batch of "good" observations which are normal $\mathcal{N}(\mu, \sigma^2)$ and "bad" ones which are normal $\mathcal{N}(\mu, 9\sigma^2)$, i.e. all observations have the same mean, but the errors of some are increased by a factor 3. Each single observation x_i is a good one with probability $1 - \varepsilon$, a bad one with probability ε, where ε is a small number. Two time-honored measures of scatter are the mean absolute deviation

$$d_n = \frac{1}{n} \sum |x_i - \bar{x}|$$

and the mean square deviation

$$s_n = \sqrt{\frac{1}{n} \sum (x_i - \bar{x})^2}.$$

There had been a dispute between Eddington and Fisher, around 1920, about the relative merits of d_n and s_n. Fisher then pointed out that for exactly normal observations, s_n is 12% more efficient than d_n, and this seemed to settle the matter.

1

Of course, the two statistics measure different characteristics of the error distribution. For instance, if the errors are exactly normal, s_n converges to σ, while d_n converges to $\sqrt{2/\pi}\sigma \approx 0.80\sigma$. So we should make precise how to compare their performance on the basis of the asymptotic relative efficiency (ARE) of d_n relative to s_n (see Table 1):

$$\mathrm{ARE}(\varepsilon) = \lim_{n\to\infty} \frac{\mathrm{var}\,(s_n)/[E(s_n)]^2}{\mathrm{var}\,(d_n)/[E(d_n)]^2}$$

$$= \frac{(3(1+80\varepsilon)/(1+8\varepsilon)^2 - 1)/4}{(\pi/2)\cdot(1+8\varepsilon)/(1+2\varepsilon)^2 - 1}.$$

TABLE 1

ε	$\mathrm{ARE}(\varepsilon)$
0	.876
.001	.948
.002	1.016
.005	1.198
.01	1.439
.02	1.752
.05	2.035
.10	1.903
.15	1.689
.25	1.371
.5	1.017
1.0	0.876

The result is disquieting: just two bad observations in 1000 suffice to offset the 12% advantage of the mean square error, and the ARE reaches a maximum value >2 at about $\varepsilon = 0.05$.

Typical "good data" samples in the physical sciences appear to be well modeled by an error law of the form

$$F(x) = (1-\varepsilon)\Phi(x) + \varepsilon\,\Phi(x/3),$$

where Φ is the standard normal cumulative, with ε in the range between 0.01 and 0.1. (This does not necessarily imply that these samples contain between 1% and 10% gross errors, although this is often true—the above may just be a convenient description of a slightly longer-tailed than normal distribution.) In other words, the naturally occurring deviations from the idealized model are large enough to render meaningless the traditional asymptotic optimality theory.

To avoid misunderstandings, I should also emphasize what is *not* implied here. First, the above results do not imply that we advocate the use of the mean absolute deviation (there are still better estimates). Second, one might argue that the example is unrealistic insofar as the "bad" observations will stick out as outliers, so any conscientious statistician will do something about them before calculating the mean square error. This is besides the point: we are concerned here with the behavior of the *unmodified* classical estimates.

The example has to do with longtailedness: lengthening the tails explodes the variability of s_n (d_n is much less affected). Shortening the tails (moving a fractional mass from the tails to the central region) on the other hand produces only negligible effects on the distributions of the estimators. (Though, it may impair absolute efficiency by decreasing the asymptotic Cramer–Rao bound; but the latter is so unstable under small changes of the distribution, that it is difficult to take this effect seriously.)

Thus, for most practical purposes, "distributionally robust" and "outlier resistant" are interchangeable. Any reasonable, formal or informal, procedure for rejecting outliers will prevent the worst.

However, only the best among these rejection rules can more or less compete with other good robust estimators. Frank Hampel (1974a), (1976) has analyzed the performance of some rejection rules (followed by the mean, as an estimate of location). Rules based on the studentized range, for instance, are disastrously bad; the maximum studentized residual and Dixon's rule are only mediocre; the top group consists of a rule based on sample kurtosis, one based on Shapiro–Wilk, and a rule which simply removes all observations x_j for which $|x_j - \text{med}_i x_i| / \text{med}_i \{|x_i - \text{med}_i x_i|\}$ exceeds a predetermined constant c (e.g. $c = 5.2$, corresponding to 3.5 standard deviations). The main reason why some of the rules performed poorly is that they won't recognize outliers if two or more of them are bundled together on the same side of the sample.

Altogether, 5–10% wrong values in a data set seem to be the rule rather than the exception (Hampel (1973a)). The worst batch I have encountered so far (some 50 ancient astronomical observations) contained about 30% gross errors.

I am inclined to agree with Daniel and Wood (1971, p. 84) who prefer technical expertise to any "statistical" criterion for straight outlier rejection. But even the thus cleaned data will not exactly correspond to the idealized model, and robust procedures should be used to process them further.

One further remark on terminology: although "robust methods" are often classified together with "nonparametric" and "distribution-free" ones, they rather belong together with classical parametric statistics. Just as there, one has an idealized parametric model, but in addition one would like to make sure that the methods work well not only at the model itself, but also in a neighborhood of it. Note that the sample mean is *the* nonparametric estimate of the population mean, but it is not robust. Distribution free tests stabilize the level, but not necessarily the power. The performance of estimates derived from rank tests tends to be robust, but since it is a function of the power, not of the level of these tests, this is a fortunate accident, not intrinsically connected with distribution-freeness.

For historical notes on the subject of robustness, see the survey articles by Huber (1972), Hampel (1973a), and in particular Stigler (1973).

CHAPTER II

Qualitative and Quantitative Robustness

2. Qualitative robustness. We have already mentioned the stability or continuity principle fundamental to (distributional) robustness: a small change in the underlying distribution should cause only a small change in the performance of a statistical procedure. These notions are due to Hampel (1968), (1971).

Thus, if $X_1, \cdots, X_n$ are i.i.d. random variables with common distribution F, and $T_n = T_n(X_1, \cdots, X_n)$ is an estimate based on them, then this requirement can be interpreted: a sufficiently small change in $F = \mathscr{L}(X)$ should result in an arbitrarily small change in $\mathscr{L}(T_n)$.

A little thought will show—if we exclude extremely pathological functions T_n—that all ordinary estimates satisfy this requirement for every fixed n. But for "nonrobust" statistics the modulus of continuity typically gets worse for increasing n. In other words, we should require that the continuity is uniform with respect to n. More precisely, for a suitable metric d_* (see below) in the space of probability measures, we require that for all $\varepsilon > 0$ there is a $\delta > 0$ such that for all $n \geqq n_0$,

$$d_*(F, G) < \delta \to d_*(\mathscr{L}_F(T_n), \mathscr{L}_G(T_n)) < \varepsilon.$$

We may require this either only at the idealized model F, or for all F, or even uniformly in F.

This situation is quite analogous to the stability of ordinary differential equations, where the solution should depend continuously (in the topology of uniform convergence) on the initial values.

However, it is somewhat tricky to work with the above formalization of robustness, and we shall begin with a much simpler, but more or less equivalent nonstochastic version of the continuity requirement.

Restrict attention to statistics which can be written as a functional T of the empirical distribution function F_n, or, perhaps better, of the empirical measure

$$\mu_n = \frac{1}{n} \sum_1^n \delta_{x_i}$$

where $(x_1, \cdots, x_n)$ is the sample and δ_x is the unit pointmass at x. Then a small change in the sample should result in a small change in the value $T_n = T(\mu_n)$ of the statistics.

Note that many of the classical statistical procedures are of this form, e.g. (i) all maximum likelihood estimates:

$$\int \log f(x, \theta) F_n(dx) = \max!,$$

(ii) linear combinations of order statistics, e.g. the α-trimmed mean

$$\bar{x}_\alpha = \frac{1}{1-2\alpha} \int_\alpha^{1-\alpha} F_n^{-1}(t)\, dt,$$

or also (iii) estimates derived from rank tests, e.g. the Hodges–Lehmann estimate

$$T_n = \tfrac{1}{2} \operatorname{med}\{F_n * F_n\}$$

(the median of the pairwise means $(x_i + x_j)/2$, with (i, j) ranging over *all* n^2 pairs; the more conventional versions, which use the pairs with $i < j$, or $i \leqq j$, are not functionals of F_n, but all three versions are of course asymptotically equivalent).

If in particular T is a linear functional, i.e. if

$$T((1-s)\mu + s\nu) = (1-s)T(\mu) + sT(\nu)$$

for $0 \leqq s \leqq 1$, then T_n can be written as

$$T_n = T(\mu_n) = \frac{1}{n}\sum T(\delta_{x_i}) = \frac{1}{n}\sum \psi(x_i) = \int \psi\, d\mu_n$$

with $\psi(x) = T(\delta_x)$.

Since we shall mostly be concerned with distributions on the real line, we shall discontinue the notational distinction between distribution functions F and measures μ and shall both denote by the same capital latin letters, with the convention that $F(\cdot)$ denotes the distribution function, $F\{\cdot\}$ the associated set function: $F(x) = F\{(-\infty, x)\}$.

A *small change* in the sample now is taken to mean either a small change in many or all of the observations ("round-off", "grouping") or a large change in a few of them ("gross errors"). For a linear functional, this means that ψ should be continuous and bounded: the first is necessary and sufficient that a small change in many x_i induces only a small change in $(1/n)\sum \psi(x_i)$; the second, that a large change in a few x_i (i.e. in a small percentage of the total sample size) induces only a small change.

Now this corresponds to the weak(-star) topology in the space of all probability measures: the weakest topology such that

$$F \to \int \psi\, dF$$

is continuous for all bounded continuous functions ψ. The weak topology thus appears to be the natural topology for statistics.

For general, not necessarily linear functionals T this naturally leads to the following basic continuity, stability or "robustness" requirement: T should be continuous with respect to the weak topology (at least at the model distribution, but if possible for all F).

For several, both technical and intuitive reasons, it is convenient to metrize the weak topology. The conceptually most attractive version is the *Prohorov metric*: let $(\Omega, \mathcal{B})$ be a complete, separable metric space with metric d, and equipped with its Borel-σ-algebra $\mathcal{B}$.

For any $A \subset \Omega$, define the closed δ-neighborhood of A as

$$A^\delta = \{x \in \Omega | \inf_{y \in A} d(x, y) \leq \delta\}.$$

It is easy to show that A^δ is closed; in fact

$$A^\delta = \bar{A}^\delta = \overline{A^\delta} = \overline{\bar{A}^\delta}.$$

Let $\mathcal{M}$ be the set of all probability measures on $(\Omega, \mathcal{B})$, let $G \in \mathcal{M}$ and let δ, $\varepsilon > 0$. Then

$$\mathcal{P}_{\delta,\varepsilon} = \{F \in \mathcal{M} | F\{A\} \leq G\{A^\delta\} + \varepsilon, \forall A \in \mathcal{B}\}$$

shall be called a *Prohorov neighborhood* of G. These neighborhoods generate the weak topology in $\mathcal{M}$.

The *Prohorov distance* is defined as

$$d_{\mathrm{Pr}}(F, G) = \inf\{\varepsilon > 0 | \forall B, F\{B\} \leq G\{B^\varepsilon\} + \varepsilon\}.$$

It is straightforward to check that d_{Pr} is a metric.

THEOREM (Strassen 1965)). *The following two statements are equivalent*:
(i) *For all $A \in \mathcal{B}$, $F\{A\} \leq G\{A^\delta\} + \varepsilon$.*
(ii) *There exist (dependent) random variables X, Y with values in Ω, such that $\mathcal{L}(X) = F$, $\mathcal{L}(Y) = G$ and $P\{d(X, Y) \leq \delta\} \geq 1 - \varepsilon$.*

In other words, if G is the idealized model and F is the true underlying distribution, such that $d_{\mathrm{Pr}}(F, G) \leq \varepsilon$, we can always assume that there is an ideal (but unobservable) random variable Y with $\mathcal{L}(Y) = G$, and an observable X with $\mathcal{L}(X) = F$, such that $P\{d(X, Y) \leq \varepsilon\} \geq 1 - \varepsilon$. That is, the model provides both for small errors occurring with large probability and large errors occurring with low probability, in a very explicit and quantitative fashion.

There are several other metrics also defining the weak topology. An interesting one is the so-called *bounded Lipschitz metric* d_{BL}. Assume that the metric on $(\Omega, \mathcal{B})$ is bounded by 1 (if necessary, replace the original metric by $d(x, y)/(1 + d(x, y))$). Then define

$$d_{\mathrm{BL}}(F, G) = \sup \left| \int \psi \, dF - \int \psi \, dG \right|$$

where ψ ranges over all functions satisfying the Lipschitz condition $|\psi(x) - \psi(y)| \leq d(x, y)$.

Also for this metric an analogue of Strassen's theorem holds (first proved in a special case by Kantorovitch and Rubinstein, 1958): $d_{\mathrm{BL}}(F, G) \leq \varepsilon$ iff there are two random variables X, Y such that $\mathcal{L}(X) = F$, $\mathcal{L}(Y) = G$, and $E \, d(X, Y) \leq \varepsilon$.

Furthermore, on the real line also the *Lévy metric* d_{L} generates the weak topology; by definition $d_{\mathrm{L}}(F, G) \leq \varepsilon$ iff

$$G(x - \varepsilon) - \varepsilon \leq F(x) \leq G(x + \varepsilon) + \varepsilon \quad \text{for all } x.$$

The Lévy metric is easier to handle than the two other ones, but unfortunately, it does not possess an intuitive interpretation in the style of the Prohorov or bounded Lipschitz metric.

It is now fairly straightforward to show that the two definitions of qualitative robustness are essentially equivalent:

THEOREM (Hampel (1971)). *Let T be defined everywhere in $\mathcal{M}$ and put $T_n = T(F_n)$. We say that T_n is consistent at F if T_n tends to $T(F)$ in probability, where F is the true underlying distribution.*

  (i) *If T is weakly continuous at all F, then T_n is consistent at all F, and $F \to \mathcal{L}(T_n)$ is weakly continuous uniformly in n.*
  (ii) *If T_n is consistent and $F \to \mathcal{L}(T_n)$ is weakly continuous uniformly in n at all F, then T is weakly continuous.*

3. Quantitative robustness, breakdown. Consider a sequence of estimates generated by a statistical functional, $T_n = T(F_n)$, where F_n is the empirical distribution. Assume that $T(F_0)$ is the target value of these estimates (the value of the functional at the idealized model distribution F_0).

Assume that the true underlying distribution F lies anywhere in some ε-neighborhood of F_0, say in

$$\mathcal{P}_\varepsilon = \{F | d_L(F_0, F) \leqq \varepsilon\}.$$

Ordinarily, our estimates will be consistent in the sense that

$$T_n \to T(F) \text{ in probability,}$$

and asymptotically normal

$$\mathcal{L}_F(\sqrt{n}(T_n - T(F))) \to \mathcal{N}(0, A(F, T)).$$

Thus, it will be convenient to discuss the quantitative asymptotic robustness properties of T in terms of the maximum bias

$$b_1(\varepsilon) = \sup_{F \in \mathcal{P}_\varepsilon} |T(F) - T(F_0)|,$$

and the maximum variance

$$v_1(\varepsilon) = \sup_{F \in \mathcal{P}_\varepsilon} A(F, T).$$

However, this is, strictly speaking, inadequate: we should like to establish that for sufficiently large n our estimate T_n behaves well for all $F \in \mathcal{P}_\varepsilon$. A description in terms of b_1 and v_1 would only allow us to show that for each $F \in \mathcal{P}_\varepsilon$, T_n behaves well for sufficiently large n. The distinction is fundamental, but has been largely neglected in the literature.

A better approach would be as follows. Let $M(F, T_n)$ be the median of $\mathcal{L}_F(T_n - T(F_0))$ and let $Q_t(F, T_n)$ be a normalized t-quantile range of $\mathcal{L}_F(\sqrt{n}T_n)$. For any distribution G, we define the normalized t-quantile range as

$$Q_t = \frac{G^{-1}(1-t) - G^{-1}(t)}{\Phi^{-1}(1-t) - \Phi^{-1}(t)}.$$

The value of t is arbitrary, but fixed, say $t = 0.25$ (interquartile range) or $t = 0.025$ (95%-range, which is convenient in view of the traditional 95% confidence intervals). For a normal distribution, Q_t coincides with the standard deviation; Q_t^2 shall also be called pseudo-variance.

Then we define the maximum asymptotic bias and variance respectively as

$$b(\varepsilon) = \lim_{n} \sup_{F \in \mathcal{P}_\varepsilon} |M(F, T_n)| \geq b_1(\varepsilon),$$

$$v(\varepsilon) = \lim_{n} \sup_{F \in \mathcal{P}_\varepsilon} Q_t(F, T_n)^2 \geq v_1(\varepsilon).$$

The inequalities here are straightforward and easy to establish, assuming that b_1 and v_1 are well defined. Since b and v are awkward to handle, we shall work with b_1 and v_1, but we are then obliged to check whether for the particular T under consideration $b_1 = b$ and $v_1 = v$. Fortunately, this is usually true.

We define the asymptotic breakdown point of T at F_0 as

$$\varepsilon^* = \varepsilon^*(F_0, T) = \sup \{\varepsilon \,|\, b(\varepsilon) < b(1)\}.$$

Roughly speaking, the breakdown point gives the maximum fraction of bad outliers the estimator can cope with. In many cases, it does not depend on F_0, nor on the particular choice of $\mathcal{P}_\varepsilon$ (in terms of Lévy distance, Prohorov distance, ε-contamination etc.)

Example. The breakdown point of the α-trimmed mean is $\varepsilon^* = \alpha$. (This is intuitively obvious; for a formal derivation see § 6.)

4. Infinitesimal robustness, influence function. For the following we assume that d_* is a metric in the space $\mathcal{M}$ of all probability measures, generating the weak topology, and which is also compatible with the affine structure of $\mathcal{M}$ in the sense that

$$d_*(F_s, F_t) = O(|s - t|),$$

where

$$F_t = (1 - t)F_0 + tF_1, \qquad 0 \leq t \leq 1.$$

We say that a statistical functional T is *Fréchet differentiable* at F if it can be approximated by a linear functional L (depending on F) such that for all G

$$|T(G) - T(F) - L(G - F)| = o(d_*(F, G)).$$

It is easy to see that L is uniquely determined: the difference L_1 of any two such functionals satisfies

$$|L_1(G - F)| = o(d_*(F, G)),$$

and in particular, with $F_t = (1 - t)F + tG$, we obtain

$$|L_1(F_t - F)| = t|L_1(G - F)| = o(d_*(F, F_t)) = o(t),$$

hence $L_1(G - F) = 0$ for all G.

Moreover, if T is weakly continuous, then L must be too. The only weakly continuous linear functionals are those of the form

$$L(G - F) = \int \psi(x) \, d(G - F)$$

for some bounded continuous function ψ. Evidently, ψ is determined only up to an additive constant, and we can standardize ψ such that $\int \psi \, dF = 0$, thus $L(G-F) = \int \psi \, dG$.

If $d_*(F, F_n)$ is of the stochastic order $O_p(n^{-1/2})$ (which holds for d_L, but not in general for d_{Pr} or d_{BL}), then we obtain an extremely simple proof of asymptotic normality:

$$\sqrt{n}(T(F_n) - T(F)) = \sqrt{n} \int \psi \, dF_n + \sqrt{n} \, o(d_*(F, F_n))$$

$$= \frac{1}{\sqrt{n}} \sum \psi(x_i) + o_p(1),$$

hence $\sqrt{n}(T(F_n) - T(F))$ is asymptotically normal with mean 0 and variance

$$\int \psi(x)^2 F(dx).$$

Unfortunately, we rarely have Fréchet differentiability, but the assertions just made remain valid under weaker assumptions (and more complicated proofs).

A functional T is called *Gâteaux differentiable*[1] at F, if there is a function ψ such that for all $G \in \mathcal{M}$,

$$\lim_{t \to 0} \frac{T((1-t)F + tG) - T(F)}{t} = \int \psi(x) G(dx).$$

Whenever the Fréchet derivative exists, then also the Gâteaux derivative does, and the two agree. Differentiable statistical functionals were first considered by von Mises (1937), (1947).

Evidently, $\psi(x)$ can be computed by inserting $G = \delta_x$ (point mass 1 at x) into the preceding formula, and in this last form it has a heuristically important interpretation, first pointed out by Hampel (1968):

$$IC(x; F, T) = \lim_{t \to 0} \frac{T((1-t)F + t\delta_x) - T(F)}{t}$$

gives the suitably scaled differential influence of one additional observation with value x, if the sample size $n \to \infty$. Therefore, Hampel has called it the *influence curve* (IC).

Note. There are moderately pathological cases where the influence curve exists, but not the Gâteaux derivative. For instance, the functional corresponding to the Bickel–Hodges estimate (Bickel and Hodges (1967))

$$\mathrm{med}\left\{\frac{x_{(i)} + x_{(n+1-i)}}{2}\right\}$$

has this property.

---

[1] Often, but erroneously, called "Volterra differentiable". See J. A. Reeds (1976).

If we approximate the influence curve as follows:

replace F by F_{n-1},

replace t by $1/n$,

we obtain the so-called sensitivity curve (Tukey (1970)):

$$SC_{n-1}(x) = \frac{T([(n-1)/n]F_{n-1} + (1/n)\delta_x) - T(F_{n-1})}{1/n}$$

$$= n[T_n(x_1, \cdots, x_{n-1}, x) - T_{n-1}(x_1, \cdots, x_{n-1})].$$

However, this does not always give a feasible approximation to the influence curve (the problem resides with the substitution of F_{n-1} for F).

If the Fréchet derivative of T at F_0 exists, then we have for the gross error model $\mathscr{P}_\varepsilon = \{F | F = (1-\varepsilon)F_0 + \varepsilon H, H \in \mathcal{M}\}$:

$$T(F) - T(F_0) = \int IC(x; F_0, T)\, dF + o(\varepsilon)$$

$$= \varepsilon \int IC(x; F_0, T)\, dH + o(\varepsilon),$$

in which case

$$b(\varepsilon) = b_1(\varepsilon) = \varepsilon \cdot \gamma^* + o(\varepsilon)$$

with

$$\gamma^* = \sup_x |IC(x; F_0, T)|.$$

γ^* has been called *gross error sensitivity* by Hampel. If we have only Gâteaux differentiability, some care is needed. We shall later give two examples where
(i) $\gamma^* < \infty$ but $b_1(\varepsilon) \equiv \infty$ for $\varepsilon > 0$,
(ii) $\gamma^* = \infty$ but $\lim b(\varepsilon) = 0$ for $\varepsilon \to 0$.

CHAPTER III

M-, L-, and R-estimates

5. M-estimates. Any estimate T_n defined by a minimum problem of the form

(5.1) $$\sum_{1}^{n} \rho(x_i; T_n) = \min!$$

or by an implicit equation

(5.2) $$\sum_{1}^{n} \psi(x_i; T_n) = 0,$$

where ρ is an arbitrary function, $\psi(x, \theta) = (\partial/\partial\theta)\rho(x; \theta)$, shall be called an *M-estimate* (or maximum likelihood type estimate; note that $\rho(x; \theta) = -\log f(x; \theta)$ gives the ordinary M.L.-estimate).

We are particularly interested in location estimates

$$\sum \rho(x_i - T_n) = \min!$$

or

$$\sum \psi(x_i - T_n) = 0.$$

If we write the last equation as

(5.3) $$\sum w_i \cdot (x_i - T_n) = 0$$

with

$$w_i = \frac{\psi(x_i - T_n)}{x_i - T_n}$$

we obtain a representation of T_n as a weighted mean

$$T_n = \frac{\sum w_i x_i}{\sum w_i}$$

with weights depending on the sample. Our favorite choices will be of the form

$$\rho(x) = x^2/2 \qquad \text{for } |x| \leq c,$$
$$= c|x| - c^2/2 \quad \text{for } |x| > c,$$
$$\psi(x) = [x]_{-c}^{c} = -c \quad \text{for } x < -c,$$
$$= x \qquad \text{for } -c \leq x \leq c,$$
$$= c \qquad \text{for } x > c,$$

13

leading to weights

$$w_i = 1 \qquad \text{for } |x_i - T_n| \le c,$$

$$= \frac{c}{|x_i - T_n|} \quad \text{for } |x_i - T_n| > c.$$

All three versions (5.1), (5.2), (5.3) are essentially equivalent. Note that the functional version of the first form

$$\int \rho(x - T(F))F(dx) = \min!$$

may cause trouble. For instance the median corresponds to $\rho(x) = |x|$, and

$$\int |x - T|F(dx) \equiv \infty$$

unless F has a finite first absolute moment. This is o.k. if we examine instead

$$\int (|x - T| - |x|)F(dx) = \min!.$$

Influence curve of an M-estimate. Put

$$F_t = (1 - t)F_0 + tF_1;$$

then the influence curve $\text{IC}(x; F_0, T)$ is the ordinary derivative

$$\dot{T} = \left[\frac{d}{dt} T(F_t) \right]_{t=0} \quad \text{with } F_1 = \delta_x.$$

In particular, for an M-estimate, i.e. for the functional $T(F)$ defined by

$$\int \psi(x; T(F))F(dx) = 0$$

we obtain by inserting F_t for F and taking the derivative (with $\psi'(x, \theta) = (\partial/\partial\theta)\psi(x, \theta)$):

$$\int \psi(x; T(F_0))d(F_1 - F_0) + \dot{T} \int \psi'(x; T(F_0))F_0(dx) = 0$$

or

$$\dot{T} = \frac{\int \psi(x; T(F_0))F_1(dx)}{-\int \psi'(x; T(F_0))F_0(dx)}.$$

After putting $F_1 = \delta_x$, we obtain

$$\text{IC}(x; F_0, T) = \frac{\psi(x; T(F_0))}{-\int \psi'(x; T(F_0))F_0(dx)}.$$

So let us remember that the influence curve of an M-estimate is simply proportional to ψ.

Breakdown and continuity properties of M-estimates. Take the location case, with T(F) defined by

$$\int \psi(x - T(F))F(dx) = 0.$$

Assume that ψ is nondecreasing, but not necessarily continuous. Then

$$\lambda(t, F) = \int \psi(x - t)F(dx)$$

is decreasing in t, increasing in F.

$T(F)$ is not necessarily unique; we have $T^* \leq T(F) \leq T^{**}$ with

$$T^* = \sup\{t | \lambda(t, F) > 0\},$$

$$T^{**} = \inf\{t | \lambda(t, F) < 0\}.$$

Now let F range over all densities with $d_L(F_0, F) \leq \varepsilon$.

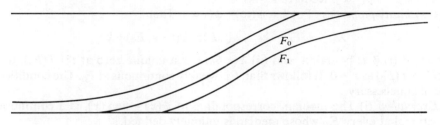

FIG. 1

The stochastically largest member of this set is the improper distribution F_1 (it puts mass ε at $+\infty$):

$$F_1(x) = F_0(x - \varepsilon) - \varepsilon \quad \text{for } x > x_0 + \varepsilon,$$

$$= 0 \qquad\qquad \text{for } x \leq x_0 + \varepsilon,$$

where x_0 is defined by

$$F_0(x_0) = \varepsilon.$$

Thus

$$\lambda(t, F) \leq \lambda(t, F_1) = \int_{x_0}^{\infty} \psi(x - t + \varepsilon)F_0(dx) + \varepsilon\psi(\infty).$$

Define

$$b_+(\varepsilon) = \sup\{T(F) | d_L(F_0, F) \leq \varepsilon\} = \inf\{t | \lambda(t, F_1) < 0\},$$

$$b_-(\varepsilon) = \inf\{T(F) | d_L(F_0, F) \leq \varepsilon\} = \sup\{t | \lambda(t, F_1) > 0\};$$

then the maximum asymptotic bias is

$$b_1(\varepsilon) = \max\{b_+(\varepsilon) - T(F_0), T(F_0) - b_-(\varepsilon)\}.$$

Breakdown. $b_+(\varepsilon) < b_+(1) = \infty$ holds iff

$$\psi(\infty) < \infty$$

and

$$\lim_{t \to \infty} \lambda(t, F_1) = (1 - \varepsilon)\psi(-\infty) + \varepsilon\psi(+\infty) < 0.$$

For $b_-(\varepsilon)$ a similar relation holds, with the roles of $+\infty$ and $-\infty$ interchanged. It follows that the breakdown point is

$$\varepsilon^* = \frac{\eta}{1 + \eta}$$

with

$$\eta = \min\left\{-\frac{\psi(-\infty)}{\psi(+\infty)}, -\frac{\psi(+\infty)}{\psi(-\infty)}\right\}.$$

If $\psi(+\infty) = -\psi(-\infty)$, we have $\varepsilon^* = \frac{1}{2}$.

Continuity properties. Put $k = \psi(\infty) - \psi(-\infty)$. Then

$$\lambda(t + \varepsilon, F_0) - k\varepsilon \leqq \lambda(t, F) \leqq \lambda(t - \varepsilon, F_0) + k\varepsilon.$$

Hence if (i) ψ is bounded, and (ii) $\lambda(t, F_0)$ has a unique zero at $t = T(F_0)$, then $T(F) \to T(F_0)$ as $\varepsilon \to 0$. It follows that T is weakly continuous at F_0. The conditions are also necessary.

Examples. (i) The median, corresponding to $\psi(x) = \text{sign}(x)$, is a continuous functional at every F_0, whose median is uniquely defined.

(ii) If ψ is bounded and strictly monotone, then the corresponding M-estimate is everywhere continuous.

6. L-estimates. Let M be a signed measure on $(0, 1)$. Then the functional

(6.1) $$T(F) = \int F^{-1}(t)M(dt)$$

induces the estimates

$$T_n = T(F_n) = \sum_{i=1}^{n} a_{ni}x_{(i)}$$

with

$$a_{ni} = M\left\{\left(\frac{i-1}{n}, \frac{i}{n}\right]\right\}.$$

In order that the T_n are translation invariant, the total algebraic mass of M should be 1.

As a special case, we first derive the influence function of $T_t = F^{-1}(t)$, the t-quantile. We take the derivative of

$$F_s(F_s^{-1}(t)) = t$$

at $s = 0$, with $F_s = (1-s)F_0 + sF_1$, and obtain

$$F_1(F_0^{-1}(t)) - F_0(F_0^{-1}(t)) + f_0(F_0^{-1}(t))\dot{T}_t = 0,$$

or

(6.2)
$$\text{IC}(x; F_0, T_t) = \frac{t - F_1(F_0^{-1}(t))}{f_0(F_0^{-1}(t))} \quad \text{with } F_1 = \delta_x.$$

Example 1. For the median $(t = \tfrac{1}{2})$ we have

$$\text{IC}(x; F, T_{1/2}) = -\frac{1}{2f(F^{-1}(\tfrac{1}{2}))}, \quad x < F^{-1}(\tfrac{1}{2}),$$

$$= \frac{1}{2f(F^{-1}(\tfrac{1}{2}))}, \quad x > F^{-1}(\tfrac{1}{2}).$$

The general case is now obtained by linear superposition:

(6.3)
$$\text{IC}(x; F, T) = \int \text{IC}(x; F, T_t)M(dt)$$

$$= \int \frac{t}{f(F^{-1}(t))} M(dt) - \int_{F(x)}^{1} \frac{1}{f(F^{-1}(t))} M(dt).$$

Example 2. If $T(F) = \sum \beta_i F^{-1}(t_i)$, then IC has jumps of size $\beta_i / f(F^{-1}(t_i))$ at the points $x = F^{-1}(t_i)$.

If M has a density m, then we may differentiate the expression (6.3) and obtain the more easily remembered formula

(6.4)
$$\frac{d}{dx}\text{IC}(x; F, T) = m(F(x)).$$

Example 3. The α-trimmed mean

$$T(F) = \frac{1}{1 - 2\alpha} \int_{\alpha}^{1-\alpha} F^{-1}(t)\, dt$$

has an influence curve of the form shown in Fig. 2.

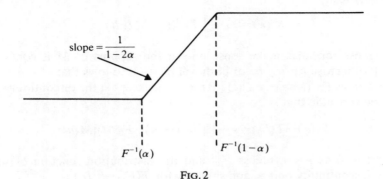

FIG. 2

Example 4. The α-Winsorized mean. For $g/n = \alpha$,

$$\bar{X}_{\alpha W} = \frac{1}{n}(gX_{(g+1)} + X_{(g+1)} + \cdots + X_{(n-g)} + gX_{(n-g)}).$$

The corresponding functional

$$\int_{\alpha}^{1-\alpha} F^{-1}(t)\,dt + \alpha(F^{-1}(\alpha) + F^{-1}(1-\alpha))$$

has the influence curve shown in Fig. 3.

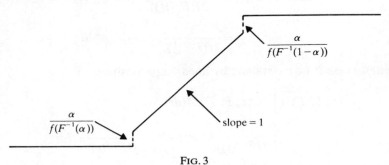

$$\frac{\alpha}{f(F^{-1}(1-\alpha))}$$

$$\frac{\alpha}{f(F^{-1}(\alpha))}$$

slope = 1

FIG. 3

Breakdown and continuity properties of L-estimates. Assume that M is a positive measure, with support contained in $[\alpha, 1-\alpha]$ where $0 < \alpha \leq \frac{1}{2}$; let α be the largest such number. As in the preceding section, we calculate

$$b_+(\varepsilon) = \sup\{T(F)|d_L(F_0, F) \leq \varepsilon\}$$

$$= \sup\left\{\int F^{-1}(t)M(dt)\right\}$$

$$= \int F_1^{-1}(t)M(dt)$$

$$= \varepsilon + \int F_0^{-1}(t+\varepsilon)M(dt),$$

and symmetrically,

$$b_-(\varepsilon) = -\varepsilon + \int F_0^{-1}(t-\varepsilon)M(dt).$$

Clearly, these formulas make sense only if the support of M is contained in $[\varepsilon, 1-\varepsilon]$; otherwise b_+ or $-b_-$ or both will be ∞. It follows that $\varepsilon^* = \alpha$.

Moreover, as $F_0^{-1}(t+\varepsilon) - F_0^{-1}(t) \downarrow 0$ for $\varepsilon \downarrow 0$, except at the discontinuity points of F_0^{-1}, we conclude that

$$b_+(\varepsilon) - T(F_0) = \varepsilon + \int [F_0^{-1}(t+\varepsilon) - F_0^{-1}(t)]M(dt)$$

converges to 0 as $\varepsilon \to 0$, unless F_0^{-1} and the distribution function of M have common discontinuity points, and similarly for $T(F_0) - b_-(\varepsilon)$.

It follows that T is continuous at all F_0 where it is well defined, i.e. at those F_0 for which F_0^{-1} and M do not have common discontinuity points.

7. R-estimates. Consider a two-sample rank test for shift: let $x_1, \cdots, x_m$ and $y_1, \cdots, y_n$ be two independent samples with distributions $F(x)$ and $G(x) = F(x - \Delta)$ respectively. Merge the two samples into one of size $m + n$, and let R_i be the rank of x_i in the combined sample. Let $a_i = a(i)$ be some given scores; then base a test of $\Delta = 0$ against $\Delta > 0$ on the test statistic

$$(7.1) \qquad S_{m,n} = \frac{1}{m} \sum_{i=1}^{m} a(R_i).$$

Usually, one assumes that the scores a_i are generated by some function J as follows:

$$(7.2) \qquad a_i = J\left(\frac{i}{m+n+1}\right).$$

But there are also other possibilities for deriving the a_i from J, and we shall prefer to work with

$$(7.3) \qquad a_i = (m+n) \int_{(i-1)/(m+n)}^{i/(m+n)} J(t)\, dt.$$

Assume for simplicity $m = n$ and put

$$(7.4) \qquad S(F, G) = \int_0^1 J(\tfrac{1}{2}s + \tfrac{1}{2}G(F^{-1}(s)))\, ds.$$

Then $S_{n,n} = S(F_n, G_n)$, where F_n, G_n are the sample distribution functions of $(x_1, \cdots, x_n)$ and $(y_1, \cdots, y_n)$ respectively, provided we define the a_i by (7.3).

One can derive estimates of shift Δ_n and location T_n from such tests:

(i) adjust Δ_n such that $S_{n,n} \approx 0$ when computed from $(x_1, \cdots, x_n)$ and $(y_1 - \Delta_n, \cdots, y_n - \Delta_n)$.

(ii) adjust T_n such that $S_{n,n} \approx 0$ when computed from $(x_1, \cdots, x_n)$ and $(2T_n - x_1, \cdots, 2T_n - x_n)$. In this case, a mirror image of the x-sample serves as a stand-in for the missing second sample.

(Note that it may not be possible to achieve an exact zero, $S_{n,n}$ being a discontinuous function.)

Example. The Wilcoxon test corresponds to $J(t) = t - \tfrac{1}{2}$ and leads to the Hodges–Lehmann estimate $T_n = \text{med}\{(x_i + x_j)/2\}$.

In terms of functionals, this means that our estimate of location derives from $T(F)$, defined by the implicit equation

$$(7.5) \qquad \int J\{\tfrac{1}{2}[s + 1 - F(2T(F) - F^{-1}(s))]\}\, ds = 0.$$

From now on we shall assume that $J(1 - t) = -J(t)$, $0 < t < 1$.

Influence curve. As in the preceding sections, we find it by inserting F_t for F into (7.5) and taking the derivative with respect to t at $t = 0$. After some calculations this gives

$$\mathrm{IC}(x; F, T) = \frac{U(x) - \int U(x)f(x)\,dx}{\int U'(x)f(x)\,dx},$$

where f is the density of F, and U is defined by its derivative

$$U'(x) = J'\{\tfrac{1}{2}[F(x) + 1 - F(2T(F) - x)]\}f(2T(F) - x).$$

If the true underlying distribution F is symmetric, there is a considerable simplification: $U(x) = J(F(x))$, and thus

$$\mathrm{IC}(x; F, T) = \frac{J(F(x))}{\int J'(F(x))f(x)^2\,dx} \quad \text{for symmetric } F.$$

Example 1. Hodges–Lehmann estimate:

$$\mathrm{IC}(x; F, T) = \frac{\tfrac{1}{2} - F(2T(F) - x)}{\int f(2T(F) - x)f(x)\,dx} \quad \text{(general } F\text{)}.$$

$$= \frac{F(x) - \tfrac{1}{2}}{\int f(x)^2\,dx} \quad \text{(symmetric } F\text{)}.$$

Example 2. Normal scores estimate $J(t) = \Phi^{-1}(t)$:

$$\mathrm{IC}(x; F, T) = \frac{\Phi^{-1}(F(x))}{\int f(x)^2/\phi[\Phi^{-1}(F(x))]\,dx} \quad \text{(symmetric } F\text{)}$$

In particular:

$$\mathrm{IC}(x; \Phi, T) = x.$$

Breakdown. The maximum bias $b_1(\varepsilon)$ and the breakdown point ε^* can be worked out as in the preceding sections if J is monotone; ε^* is that value of ε for which

$$\int_{1/2}^{1-\varepsilon/2} J(s)\,ds = \int_{1-\varepsilon/2}^{1} J(s)\,ds.$$

Hodges–Lehmann: $\varepsilon^* = 1 - 1/\sqrt{2} \approx 0.293$

Normal Scores: $\varepsilon^* = 2\Phi(-\sqrt{\ln 4}) \approx 0.239$.

Note that the normal scores estimate is robust at Φ, even though the influence curve is unbounded there.

8. Asymptotic properties of M-estimates. The influence function gives a nice, unified description of the asymptotic properties of arbitrary statistical functionals

$T(F_n)$: it is usually true that $\sqrt{n}(T(F_n) - T(F))$ is asymptotically normal with mean 0 and variance

$$\int IC(x; F, T)^2 F(dx).$$

However, proofs via the influence function are rarely viable (or only under too restrictive regularity assumptions.)

We shall now sketch a rigorous treatment of the asymptotic properties of M-estimates.

Assume that $\psi(x, t)$ is measurable in x and decreasing ($=$ nonincreasing) in t, from strictly positive to strictly negative values. Put

$$T_n^* = \sup\left\{t \,\Big|\, \sum_1^n \psi(x_i, t) > 0\right\},$$

$$T_n^{**} = \inf\left\{t \,\Big|\, \sum_1^n \psi(x_i, t) < 0\right\}.$$

We have $-\infty < T_n^* \leq T_n^{**} < \infty$, and any value $T_n^* \leq T_n \leq_n^{**}$ can serve as our estimate. Note that

$$\{T_n^* < t\} \subset \{\textstyle\sum \psi(x_i, t) \leq 0\} \subset \{T_n^* \leq t\},$$

$$\{T_n^{**} < t\} \subset \{\textstyle\sum \psi(x_i, t) < 0\} \subset \{T_n^{**} \leq t\}.$$

Hence

$$P\{T_n^* < t\} = P\{\textstyle\sum \psi(x_i, t) \leq 0\},$$

$$P\{T_n^{**} < t\} = P\{\textstyle\sum \psi(x_i, t) < 0\}$$

at the continuity points of the left hand side. For instance, the randomized estimate: $T_n = T_n^*$ or $= T_n^{**}$ with equal probability $\frac{1}{2}$ (but not the more customary midpoint estimate $T_n = (T_n^* + T_n^{**})/2$) has the explicitly expressible distribution function

$$P\{T_n < t\} = \tfrac{1}{2}P\{\textstyle\sum \psi \leq 0\} + \tfrac{1}{2}P\{\textstyle\sum \psi < 0\}.$$

It follows that the exact distributions of T_n^*, T_n^{**}, T_n can be calculated from the convolution powers of $\mathscr{L}(\psi(x, t))$. Asymptotic approximations can be found by expanding $G_n = \mathscr{L}(\sum_1^n \psi(x_i, t))$ into an asymptotic series.

We may take the traditional Edgeworth expansion

$$(8.1) \qquad G_n(x) \sim \Phi(x) + \varphi(x)\left[\frac{1}{\sqrt{n}}R_3(x) + \frac{1}{n}R_4(x) + \cdots\right],$$

but it is advantageous and gives much better results for smaller n, if we expand g_n'/g_n instead of G_n or g_n, and if we center the distributions not at 0, but at the point of interest. Thus, if we have independent random variables Y_i with density $f(x)$ and would like to determine the distribution G_n of $Y_1 + \cdots + Y_n$ at the point t, we replace the original density f by f_t

$$f_t(z) = c_t e^{a_t z} f(t + z),$$

where c_t and α_t are chosen such that this is a probability density with expectation 0. Denote its variance by

$$\sigma_t^2 = \int z^2 f_t(z)\, dz,$$

and

$$\lambda_{3,t} = \int z^3 f_t(z)\, dz / \sigma_t^3.$$

Then we get from the first two terms of the Edgeworth expansion (8.1), calculated at $x = 0$, that

$$\frac{g_n'(t)}{g_n(t)} \sim -n\alpha_t - \frac{\lambda_{3,t}}{2\sigma_t}.$$

From this we obtain g_n and G_n by two numerical integrations and an exponentiation; the integration constant must be determined such that G_n has total mass 1. It turns out that the first integration can be done explicitly:

$$\log g_n(t) \sim -nc_t - \log \sigma_t + \text{const.}$$

This variant of the saddle point method was suggested by F. Hampel (1973b) who realized that the principal approximation error was residing in the normalizing constant and that it could be avoided by expanding g_n'/g_n and then determining this constant by numerical integration. See also a forthcoming paper by H. E. Daniels (1976).

We now revert to the limit law. Put

$$\lambda(t) = E\psi(X, t).$$

Assume that $\lambda(t)$ and $E[\psi(X, t)^2]$ are continuous in t and finite, and that $\lambda(t_0) = 0$ for some t_0. Then one shows easily that

$$\mathcal{L}(\sqrt{n}\lambda(T_n)) \to \mathcal{N}(0, E[\psi(X, t_0)^2]).$$

If λ is differentiable at t_0, and if we can interchange the order of integration and differentiation, we obtain from this

$$\text{as. var } (\sqrt{n}T_n) = \frac{E[\psi(X, t_0)^2]}{[\lambda'(t_0)]^2} = \frac{E(\psi^2)}{(E\psi')^2}.$$

The last expression corresponds to what one formally gets from the influence function.

9. Asymptotically efficient M-, L-, R-estimates. Let $(F_\theta)_{\theta \in \Theta}$ be a parametric family of distributions. We shall first present a heuristic argument that a Fisher consistent estimate of θ, i.e. a functional T satisfying

$$T(F_\theta) = \theta$$

is asymptotically efficient if and only if its influence function satisfies

$$\text{IC}(x; F_\theta, T) = \frac{1}{I(F_\theta)} \frac{\partial}{\partial \theta} (\log f_\theta).$$

Here, f_θ is the density of F_θ, and

$$I(F_\theta) = \int \left(\frac{\partial}{\partial \theta} \log f_\theta \right)^2 dF_\theta$$

is the Fisher information. For Fréchet differentiable functionals the argument is rigorous.

Assume that $d_L(F_\theta, F_{\theta+\delta}) = O(\delta)$, that

$$\frac{f_{\theta+\delta} - f_\theta}{\delta \cdot f_\theta} \xrightarrow{L_2(F_\theta)} \frac{\partial}{\partial \theta} \log f_\theta$$

and that

$$0 < I(F_\theta) < \infty.$$

Then, by the definition of the Fréchet derivative

$$T(F_{\theta+\delta}) - T(F_\theta) - \int \text{IC}(x; F_\theta, T)(f_{\theta+\delta} - f_\theta) \, dx = o(d_L(F_\theta, F_{\theta+\delta})) = o(\delta).$$

We divide this by δ and let $\delta \to 0$. This gives

$$\int \text{IC}(x; F_\theta, T) \frac{\partial}{\partial \theta} (\log f_\theta) f_\theta \, dx = 1.$$

The Schwarz inequality applied to this equation first gives that the asymptotic variance $A(F_\theta, T)$ of $\sqrt{n} T(F_n)$ satisfies

$$A(F_\theta, T) = \int \text{IC}(x; F\theta, T)^2 \, dF_\theta \geqq \frac{1}{I(F_\theta)},$$

and second, that we can have equality only if $\text{IC}(x; F_\theta, T)$ is proportional to $(\partial/\partial\theta) \log (f_\theta)$. This yields the result announced initially.

For the location case, $F_\theta(x) = F_0(x - \theta)$, we thus obtain asymptotic efficiency with the following choices:

(i) M-estimate:

$$\psi(x) = -c \cdot \frac{f_0'(x)}{f_0(x)}, \qquad c \neq 0.$$

(ii) L-estimate:

$$m(F_0(x)) = -\frac{1}{I(F_0)} [\log f_0(x)]''.$$

(iii) R-estimate: if F_0 is symmetric, then

$$J(F_0(x)) = -c \frac{f_0'(x)}{f_0(x)}, \qquad c \neq 0.$$

For asymmetric F_0 one cannot achieve full efficiency with R-estimates.

Of course, one must check in each individual case whether these estimates are indeed efficient (the rather stringent regularity conditions—Fréchet differentiability—will rarely be satisfied).

Examples of asymptotically efficient estimates for different distributions follow:

Example 1. Normal distribution $f_0(x) = (1/\sqrt{2\pi})\,e^{-x^2/2}$.

M; $\psi(x) = x$ sample mean, nonrobust,

L: $m(t) = 1$ sample mean, nonrobust,

R: $J(t) = \Phi^{-1}(t)$ normal scores estimate, robust.

(The normal scores estimate loses its high efficiency very quickly when only a small amount of far-out contamination is added and is soon surpassed by the Hodges–Lehmann estimate. Thus, although it is both robust and fully efficient at the normal model, I would hesitate to recommend the normal scores estimate for practical use.)

Example 2. Logistic distribution $F_0(x) = 1/(1 + e^{-x})$.

M: $\psi(x) = \tanh(x/2)$ robust,

L: $m(t) = 6t(1-t)$ nonrobust,

R: $J(t) = t - \frac{1}{2}$ robust

(Hodges–Lehmann estimate).

Remember that an L-estimate is robust iff support $(M) = [\alpha, 1-\alpha]$ for some $0 < \alpha < \frac{1}{2}$. Note that the influence function of the logistic L-estimate satisfies

$$\frac{d}{dx}\, \mathrm{IC}(x; F, T) = 6F(x)(1 - F(x))$$

and thus is highly unstable under even small changes (for any F with Cauchy-like tails, IC becomes unbounded).

In this connection, the following trivial identity can be extremely useful:

$$T(F_1) - T(F_0) = \int_0^1 \int \mathrm{IC}(x; F_s, T)(dF_1 - dF_0)\, ds,$$

where $F_s = (1-s)F_0 + sF_1$. While the influence function is the single most important tool to assess the robustness and asymptotic behavior of an estimate, the above formula and the preceding examples show that it is not enough to know the influence curve *at* the idealized model distribution—we must also know its behavior in a *neighborhood of* the model.

Example 3.

$$f_0(x) = Ce^{-x^2/2}, \qquad |x| \le c,$$
$$= Ce^{-c|x|+c^2/2}, \quad |x| \ge c.$$

Then

$$-\frac{f_0'(x)}{f_0(x)} = [x]_{-c}^c = \max\,(-c, \min\,(c, x)).$$

The asymptotically efficient estimates are:
— the M-estimate with $\psi(x) = [x]_{-c}^c$,
— an α-trimmed mean $(\alpha = F_0(-c))$,
— a somewhat complicated R-estimate.

10. Scaling questions. Scaling problems arise in two conceptually unrelated contexts: first because M-estimates of location as hitherto defined are not scale invariant; second when one is estimating the error of a location estimate. In order to make location estimates scale invariant, one must combine them with an (equivariant) scale estimate S_n:

$$\sum \psi\!\left(\frac{x_i - T_n}{S_n}\right) = 0.$$

For straight location, the median absolute deviation (MAD) appears to be the best ancillary scale estimate:

$$S_n = \mathrm{MAD}_n = \mathrm{med}\,\{|x_i - m|\}$$

where

$$m = \mathrm{med}\,\{x_i\}.$$

The somewhat poor efficiency of MAD_n is more than counter-balanced by its high breakdown point $(\varepsilon^* = \frac{1}{2})$; also T_n retains that high breakdown point.

In more complicated situations, e.g. regression (see below), the median absolute deviation loses many of its advantages, and in particular, it is no longer an easily and independently of T_n computable statistic.

The second scaling problem is the estimation of the statistical error of T_n. The asymptotic variance of $\sqrt{n}\,T_n$ is

$$(10.1) \qquad A(F, T) = \int \mathrm{IC}(x; F, T)^2 F(dx).$$

The true underlying F is unknown: we may estimate $A(F, T)$ by substituting either F_n or F_0 (supposedly F is close to the model distribution F_0), or a combination of them, on the right hand side—provided IC depends in a sufficiently nice way on F. Otherwise we may have to resort to more complicated "smoothed" estimates of IC $(\,\cdot\,; F, T)$.

For complicated estimates $\mathrm{IC}(x; F, T)$ may not readily be computable, so we might replace it by the sensitivity curve SC_n (see § 4). If we intend to integrate with respect to $F_n(dx)$, we only need the values of SC_n at the observations x_i. Then, instead of doubling the observations at x_i, we might just as well leave it out when computing the difference quotient, i.e., we would approximate $IC(x_i)$ by

$$\frac{T[n/(n-1)F_n - 1/(n-1)\delta_{x_i}] - T(F_n)}{-1/(n-1)}$$

(10.2)
$$= (n-1)(T_n(x_1, \cdots, x_n) - T_{n-1}(x_1, \cdots, x_{i-1}, x_{i+1}, \cdots, x_n)).$$

This is related to the so-called jackknife (Miller (1964), (1974)).

Jackknife. Consider an estimate $T_n(x_1, \cdots, x_n)$ which is essentially the "same" across different sample sizes. Then the *i-th jackknifed pseudo-value* is, by definition

$$T_{ni}^* = nT_n - (n-1)T_{n-1}(x_1, \cdots, x_{i-1}, x_{i+1}, \cdots, x_n).$$

If T_n is the sample mean, then $T_{ni}^* = x_i$, for example.

In terms of the jackknife, our previous approximation (10.2) to the influence curve is $= T_{ni}^* - T_n$.

If T_n is a consistent estimate of θ whose bias has the expansion

$$E(T_n - \theta) = \frac{a_1}{n} + \frac{a_2}{n^2} + O\left(\frac{1}{n^3}\right),$$

then

$$T_n^* = \frac{1}{n}\sum_1^n T_{ni}^*$$

has a smaller bias:

$$E(T_n^* - \theta) = -\frac{a_2}{n^2} + O\left(\frac{1}{n^3}\right);$$

see Quenouille (1956).

Example. If

$$T_n = \frac{1}{n}\sum (x_i - \bar{x})^2,$$

then

$$T_{ni}^* = \frac{n}{n-1}(x_i - \bar{x})^2,$$

and

$$T_n^* = \frac{1}{n-1}\sum (x_i - \bar{x})^2.$$

Tukey (1958) pointed out that

(10.3)
$$\frac{1}{n(n-1)}\sum (T_{ni}^* - T_n^*)^2$$

usually is a good estimator of the variance of T_n. (It can also be used as an estimate of the variance of T_n^*, but it is better matched to T_n.)

Warning. If the influence function $IC(x; F, T)$ does not depend smoothly on F, for instance in the case of the sample median, the jackknife is in trouble and may yield a variance estimate which is worse than useless.

Example. The α-trimmed mean $\bar{x}_\alpha$. Assume for simplicity that $g = (n-1)\alpha$ is an integer, and that $x_1 < x_2 < \cdots < x_n$. Then

$$T^*_{ni} = \frac{1}{1-2\alpha}(x_{w(i)} - \alpha x_{g+1} - \alpha x_{n-g})$$

where

$$w(i) = g+1 \quad \text{if } i \leq g+1,$$
$$= i \qquad \text{if } g+1 < i < n-g,$$
$$= n-g \quad \text{if } i \geq n-g.$$

Note that $(x_{w(i)})_{1 \leq i \leq n}$ is a Winsorized sample. The jackknife estimate of the variance of $\bar{x}_\alpha$ now turns out to be a suitably scaled Winsorized variance:

$$(10.4) \qquad \frac{1}{n(n-1)} \frac{1}{(1-2\alpha)^2} \sum (x_{w(i)} - \bar{x}_w)^2$$

where $\bar{x}_w = (1/n) \sum x_{w(i)}$ is the Winsorized mean. If $g = (n-1)\alpha$ is not an integer, the situation is somewhat more complicated but the formula (10.4) for the variance estimate remains valid, if we define x_{g+1} by linearly interpolating between the two adjacent order statistics $x_{\lfloor g+1 \rfloor}$ and $x_{\lceil g+1 \rceil}$, and similarly for x_{n-g}. If g is an integer, then $T^*_n = T_n$ is the trimmed mean, but otherwise T^*_n is a kind of Winsorized mean.

Note that, apart from minor details like the $n-1$ in the denominator, formula (10.4) coincides with $\int IC(x, F_n, \bar{x}_\alpha)^2 \, dF_n$. Here, as well as in other cases where the influence curve is explicitly known, this last formula may be somewhat cheaper to evaluate than (10.3).

CHAPTER IV

Asymptotic Minimax Theory

11. Minimax asymptotic bias. To fix the idea, assume that the true distribution F lies in the set

$$\mathscr{P}_\varepsilon = \{F \mid F = (1-\varepsilon)\Phi + \varepsilon H, H \in \mathscr{M}\}.$$

The median incurs its maximal positive bias x_0 when all contamination lies to the right of x_0, where x_0 is determined from

$$(1-\varepsilon)\Phi(x_0) = \tfrac{1}{2};$$

i.e., for the median we obtain

$$b_1(\varepsilon) = x_0 = \Phi^{-1}\left(\frac{1}{2(1-\varepsilon)}\right).$$

On the other hand, $\mathscr{P}_\varepsilon$ contains the following distribution F_+, defined by its density

$$f_+(x) = (1-\varepsilon)\varphi(x), \qquad x \le x_0,$$
$$= (1-\varepsilon)\varphi(x - 2x_0), \quad x > x_0,$$

where φ is the standard normal density. Note that F_+ is symmetric around x_0, and that

$$F_-(x) = F_+(x + 2x_0)$$

also belongs to $\mathscr{P}_\varepsilon$. Thus, we must have

$$T(F_+) - T(F_-) = 2x_0$$

for any translation invariant functional T. It is obvious from this that none can have a smaller absolute bias than x_0 at F_+ and F_- simultaneously.

For the median, we have (rather trivially) $b_1(\varepsilon) = b(\varepsilon)$, and thus we have shown that the sample median minimizes the maximal asymptotic bias.

We did not use any particular property of $\mathscr{P}_\varepsilon$, and the same argument carries through with little changes for other distributions than the normal and other types of neighborhoods. It appears, not surprisingly, that the sample median is the estimate of choice for extremely large sample sizes, where the possible bias becomes more important than the standard deviation of the estimate (which is of the order $1/\sqrt{n}$).

12. Minimax asymptotic variance. In the following, $\mathscr{P}$ will be some neighbor-hood of the normal distribution Φ, consisting of symmetric distributions only, e.g.

$$\mathscr{P} = \{F|F = (1 - \varepsilon)\Phi + \varepsilon H, H \text{ symmetric}\},$$

or

$$\mathscr{P} = \{F|d_*(\Phi, F) \leqq \varepsilon, F \text{ symmetric}\}.$$

It is convenient to assume that $\mathscr{P}$ be convex and compact in a suitable topology (the vague topology: the weakest that $F \to \int \psi \, dF$ is continuous for all continuous ψ with compact support). We allow $\mathscr{P}$ to contain substochastic measures (i.e. probability measures putting mass at $\pm\infty$); these may be thought to formalize the possibility of infinitely bad outliers. The problem is to estimate location θ in the family $F(x - \theta)$, $F \in \mathscr{P}$.

The theory is described in some detail in Huber (1964); I only sketch the salient points here.

First, we have to minimize Fisher information over $\mathscr{P}$.

1. Define Fisher information as

$$I(F) = \sup_{\psi \in \mathscr{C}_k^1} \frac{(\int \psi' \, dF)^2}{\int \psi^2 \, dF},$$

(where $\mathscr{C}_k^1$ is the set of continuously differentiable functions with compact support).

2. THEOREM. *The following two assertions are equivalent*:
   (i) $I(F) < \infty$.
   (ii) *F has an absolutely continuous density f, and $\int (f'/f)^2 f \, dx < \infty$.*
In either case, $I(F) = \int (f'/f)^2 f \, dx$.

Proof (Liggett). Define a linear functional A by

$$A\psi = -\int \psi' \, dF.$$

Note that $I(F)$ is the square of the L_2-norm of A, hence A is bounded if $I(F)$ is finite. By Riesz' theorem then there is a $g \in L_2(F)$ such that

$$A\psi = \int \psi g \, dF \quad \text{for all } \psi \in L_2(F).$$

We do not know yet whether F has an absolutely continuous density, but if it has, then

$$A\psi = -\int \psi' f \, dx = \int \psi \frac{f'}{f} f \, dx,$$

hence $g = f'/f$. We thus *define*

$$f(x) = \int_{y<x} g(y) F(dy).$$

The rest of the proof is a matter of straightforward verification.

3. $I(\cdot)$ is lower semi-continuous (being the supremum of a set of continuous functions); hence it attains its minimum on the compact set $\mathcal{P}$, say at F_0.

4. $I(\cdot)$ is convex.

5. If $f_0 > 0$, then F_0 is unique.

6. The formal expression for the inverse of the asymptotic variance of an M-estimate of location,

$$\frac{1}{A(F, \psi_0)} = \frac{(\int \psi_0' \, dF)^2}{\int \psi_0^2 \, dF}$$

is convex in F.

7. Take $\psi_0 = -f_0'/f_0$. Then

$$\frac{1}{A(F_0, \psi_0)} = I(F_0).$$

8. Let $F_t = (1-t)F_0 + tF_1$, $I(F_1) < \infty$.
Then, explicit calculation gives

$$\left[\frac{d}{dt} \frac{1}{A(F_t, \psi_0)}\right]_{t=0} = \left[\frac{d}{dt} I(F_t)\right]_{t=0} \geq 0.$$

It follows that the asymptotic variance $A(F, \psi_0)$ attains its maximum over $\mathcal{P}$ at F_0, but there the M-estimate is asymptotically efficient.

Hence the M.L.-estimate for location based on F_0 is minimax over $\mathcal{P}$.

Example. For ε-contamination, we obtain

$$f_0(x) = \frac{1-\varepsilon}{\sqrt{2\pi}} e^{-\rho(x)}$$

with

$$\rho(x) = \frac{x^2}{2}, \qquad |x| \leq c,$$

$$= c|x| - \frac{c^2}{2}, \qquad |x| > c,$$

where $c = c(\varepsilon)$.

The L- and R-estimates which are efficient at F_0 do not necessarily yield minimax solutions, since convexity fails (point 6 in the above sketch of the proof). There are in fact counter-examples (Sacks and Ylvisaker (1972)). However, in the important case of symmetric ε-contamination, the conclusion remains true for both L- and R-estimates (Jaeckel (1971a)).

Variants. Note that the least informative distribution F_0 has exponential tails, i.e. they might be *slimmer*(!) than what one would expect in practice. So it might be worthwhile to increase the maximum risk a little bit beyond the minimax value in order to gain a better performance at long tailed distributions.

This can be done as follows. Consider M-estimates, and minimize the maximal asymptotic variance subject to the side condition

$$\psi(x) = 0 \quad \text{for } |x| > q.$$

The solution for contaminated normal distributions is of the form (Collins (1976)) (see Fig. 4):

$$\psi(x) = x, \qquad 0 \le x \le c,$$
$$\psi(x) = b \tanh\left[\tfrac{1}{2}b(q-x)\right], \quad c \le x \le q$$
$$= 0 \qquad\qquad x \ge q,$$
$$\psi(-x) = -\psi(x).$$

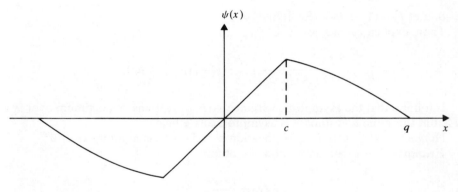

FIG. 4

The values of c and b of course depend on ε.

The actual performance does not depend very much on exactly how ψ redescends to 0, only one should make sure that it does not do it too steeply; in particular $|\psi'|$ should be small when $|\psi|$ is large.

Hampel's extremal problem (Hampel (1968)). Assume that the model is a general one-parameter family of densities $f(x, \theta)$ and estimate θ by an M-estimate based on some function $\psi(x, \theta)$, i.e.

$$\sum_1^n \psi(x_i, T_n) = 0.$$

Assume that T is "Fisher consistent" at the model, i.e.

$$\int \psi(x, \theta) f(x, \theta)\, dx = 0.$$

Then

$$\mathrm{IC}(x\,; F_\theta,\, T) = -\frac{\psi(x,\,\theta)}{\int (\partial/\partial\theta)\psi(x,\,\theta)f(x,\,\theta)\,dx}$$

$$= \frac{\psi(x,\,\theta)}{\int \psi(x,\,\theta)(\partial/\partial\theta)f(x,\,\theta)\,dx}$$

and the asymptotic variance at the model is

$$A\,(F_\theta,\, \psi) = \frac{\int \psi(x,\,\theta)^2 f(x,\,\theta)\,dx}{(\int \psi(x,\,\theta)(\partial/\partial\theta)f(x,\,\theta)\,dx)^2}.$$

Hampel's extremal problem now is to put a bound on the gross error sensitivity:

$$\sup_x |\mathrm{IC}(x\,; F_\theta,\, T)| \leqq k_\theta$$

with some appropriately chosen function k_θ and, subject to this side condition, to minimize the asymptotic variance $A\,(F_\theta,\, \psi)$ *at the model*.

The solution is of the form

$$\psi(x,\,\theta) = \left[\frac{(\partial/\partial\theta)f(x,\,\theta)}{f(x,\,\theta)} - a_\theta\right]_{-b_\theta}^{b_\theta},$$

where we have used the notation

$$[x]_a^b = a \quad \text{for } x \leqq a,$$
$$= x \quad \text{for } a < x < b,$$
$$= b \quad \text{for } x \geqq b.$$

The functions a_θ, b_θ are somewhat difficult to determine, and if k_θ is too small, there is no solution at all. It might therefore be preferable to start with choosing b_θ. A reasonable choice might be

$$b_\theta = c\sqrt{I(F_\theta)}$$

with c between 1 and 2, and where

$$I(F_\theta) = \int \left(\frac{\partial}{\partial\theta} \log f(x,\,\theta)\right)^2 f(x,\,\theta)\,dx$$

is the Fisher information. Then one determines a_θ (so that the estimate is Fisher consistent), and finally, one finds k_θ.

CHAPTER V

Multiparameter Problems

13. Generalities. As far as M-estimates are concerned, most concepts of the preceding chapters generalize to vector valued parameters. Asymptotic normality was treated by Huber (1967). The Fisher information matrix and the inverse of the asymptotic covariance matrix of the estimate are convex functions of the true underlying distribution, matrices being ordered by positive definiteness. Since this is not a lattice ordering, it is not in general possible to find a distribution minimizing Fisher information. But if there is one, the corresponding maximum likelihood estimate possesses an asymptotic minimax property: it minimizes the maximal asymptotic variance among all distributions for which it is Fisher consistent.

14. Regression. Assume that p unknown parameters $(\theta_1, \cdots, \theta_p) = \boldsymbol{\theta}^T$ are to be estimated from n observations $(y_1, \cdots, y_n) = \mathbf{y}^T$, to which they are related by

$$(14.1) \qquad\qquad y_i = f_i(\boldsymbol{\theta}) + u_i.$$

The f_i are known functions, often assumed to be linear in $\boldsymbol{\theta}$, and the u_i are independent random errors with approximately identical distributions.

One wants to estimate the unknown true $\boldsymbol{\theta}$ by a value $\hat{\boldsymbol{\theta}}$ such that the residuals

$$(14.2) \qquad\qquad \Delta_i = \Delta_i(\boldsymbol{\theta}) = y_i - f_i(\boldsymbol{\theta})$$

are made "as small as possible".

Classically, this is interpreted (Gauss, Legendre) as

$$(14.3) \qquad\qquad \sum_i \Delta_i^2 = \min!,$$

or, almost equivalently, by taking derivatives:

$$(14.4) \qquad\qquad \sum_i \Delta_i \frac{\partial f_i}{\partial \theta_j} = 0, \qquad\qquad j = 1, \cdots, p.$$

Unfortunately, this classical approach is highly sensitive to occasional gross errors. As a remedy, one may replace the square in (14.3) by a less rapidly increasing function ρ:

$$(14.5) \qquad\qquad \sum \rho(\Delta_i) = \min!,$$

35

or, instead of (14.4), to solve

$$(14.6) \qquad \sum_i \psi(\Delta_i) \frac{\partial f_i}{\partial \theta_j} = 0, \qquad\qquad j = 1, \cdots, p,$$

for $\boldsymbol{\theta}$ with $\psi = \rho'$.

There are also other possibilities to robustify (14.3). For instance, Jurečková (1971) and Jaeckel (1972) have proposed to replace the residuals Δ_i in (14.4) by their ranks (or more generally, by a function of their ranks). Possibly, it might be good to safeguard against errors in the f_i by modifying also the second factor in (14.4), e.g. by replacing also $\partial f_i / \partial \theta_j$ by its rank in $(\partial f_1 / \partial \theta_j, \cdots, \partial f_n / \partial \theta_j)$, but the consequences of these ideas have only begun to be investigated (Hill (1977)).

In any case, the empirical evidence available to date suggests that the M-estimate approach (14.5), (14.6) is easier to handle and more flexible, and even has slightly better statistical properties than the approaches based on R- and L-estimates. There is only one minor disadvantage: one must simultaneously estimate a scale parameter S in order to make it scale invariant, e.g.

$$\sum \rho\left(\frac{\Delta_i}{S}\right) = \min!$$

where S is determined simultaneously from

$$\frac{1}{n-p} \sum \psi\left(\frac{\Delta_i}{S}\right)^2 = \beta = E_\Phi(\psi(x)^2).$$

In the regression case I would prefer this S to, say, the median absolute value of the residuals, since it is more easily tractable in theory (convergence proofs) and since it fits better into the established flow of calculation of $\hat\theta$ in large least squares problems.

In order that robust regression works, the observation y_i should not have an overriding influence on the fitted value

$$\hat{y}_i = f_i(\hat{\boldsymbol{\theta}}).$$

To clarify the issues, take the classical least squares case and assume the f_i to be linear:

$$f_i(\boldsymbol{\theta}) = \sum c_{ij}\theta_j.$$

Then

$$\hat{\mathbf{y}} = \Gamma \mathbf{y}$$

with

$$\Gamma = C(C^T C)^{-1} C^T.$$

If var $(y_i) = \sigma^2$, we obtain

$$\operatorname{var}(\hat{y}_i) = \gamma_{ii}\sigma^2$$

and

$$\operatorname{var}(\Delta_i) = \operatorname{var}(y_i - \hat{y}_i) = (1 - \gamma_{ii})\sigma^2,$$

where γ_{ii} is the ith diagonal element of Γ.

Note that $\text{tr}\,(\Gamma) = p$, so $\max \gamma_{ii} \geqq \text{ave } \gamma_{ii} = p/n$; in some sense, $1/\gamma_{ii}$ is the effective number of observations entering into the determination of $\hat{y}_i$. If γ_{ii} is close to 1, $\hat{y}_i$ is essentially determined by y_i alone, y_i may have an undue leverage on the determination of certain parameters, and it may well be impossible to decide whether y_i contains a gross error or not.

The asymptotic theory of robust regression works if $\varepsilon = \max \gamma_{ii}$ goes to 0 sufficiently fast when p and n tend to infinity; "sufficiently fast" may be taken to mean $\varepsilon p^2 \to 0$ or (with slightly weaker results) $\varepsilon p \to 0$.

If only $\varepsilon \to 0$, there may be trouble if the observational errors have an asymmetric distribution and p is extremely large (above 100). This effect has been experimentally verified in a specifically designed Monte Carlo study (Huber (1973a)), but for all practical proposes, $\varepsilon \to 0$ seems to be o.k. Note that $\varepsilon \to 0$ implies $p/n \to 0$.

As already mentioned, we propose to enforce scale invariance by estimating a scale parameter σ simultaneously with $\boldsymbol{\theta}$. This can be done elegantly by minimizing an expression of the form

$$Q(\boldsymbol{\theta}, \sigma) = \sum \rho\!\left(\frac{y_i - f_i(\boldsymbol{\theta})}{\sigma}\right)\sigma + a\sigma, \qquad \sigma > 0.$$

(The more natural looking expression derived from the simultaneous M.L.-problem, which contains $\log \sigma$, would not allow us to push through a simple convergence proof for the numerical calculations.)

In the above, $\rho \geqq 0$ is a convex function, $\rho(0) = 0$, which should satisfy

$$0 < \lim_{|x| \to \infty} \frac{\rho(x)}{|x|} = c \leqq \infty.$$

If $c < \infty$, Q can be extended by continuity

$$Q(\theta, 0) = c \sum |y_i - f_i(\boldsymbol{\theta})|.$$

One easily checks that Q is a convex function of $(\boldsymbol{\theta}, \sigma)$ if the f_i are linear. Unless the minimum $(\hat{\boldsymbol{\theta}}, \hat{\sigma})$ occurs on the boundary $\sigma = 0$, it can equivalently be characterized by the $p + 1$ equations

$$\sum \psi\!\left(\frac{\Delta_i}{\hat{\sigma}}\right)\frac{\partial f_i}{\partial \theta_j} = 0, \qquad\qquad j = 1, \cdots, p,$$

$$\sum \chi\!\left(\frac{\Delta_i}{\hat{\sigma}}\right) \quad = a,$$

with

$$\Delta_i = y_i - f_i(\hat{\boldsymbol{\theta}}),$$
$$\psi(x) = \rho'(x),$$
$$\chi(x) = x\psi(x) - \rho(x).$$

If $\hat{\sigma}$ is to be asymptotically unbiased for normal errors, we should choose

$$a = (n-p)E\chi(U)$$

where the expectation is taken for a standard normal argument U.

Examples.

(i) With $\rho(x) = x^2/2$ we obtain the standard least squares estimates: $\hat{\theta}$ minimizes (14.3) and $\hat{\sigma}$ satisfies

$$\hat{\sigma}^2 = \frac{1}{n-p} \sum \Delta_i^2.$$

(ii) With

$$\rho(x) = \frac{x^2}{2}, \qquad |x| \leqq c,$$

$$= c|x| - \frac{c^2}{2}, \quad |x| > c,$$

we have

$$\chi = \tfrac{1}{2}\psi^2,$$

and we obtain the "proposal 2"-estimates of Huber (1964), (1973).

Algorithms. I know essentially three successful algorithms for minimizing Q. The simultaneous estimation of σ introduces some (inessential) complications, and if we disregard it for the moment, we may describe the salient ideas as follows.

ALGORITHM S. Apply Newton's method to $\sum \psi(\Delta_i)\partial f_i/\partial\theta_j = 0$, $j = 1, \cdots, p$. If ψ is piecewise linear, if the f_i are linear, and if the trial value $\theta^{(m)}$ is so close to the final value $\hat{\theta}$ that both induce the same classification of the residuals Δ_i according to the linear pieces of ψ they lie in, then this procedure reaches the *exact*, final value $\hat{\theta}$ in one single step. If the iteration step for $\sigma^{(m)}$ is arranged properly, this also holds for the scale invariant version of the algorithm.

ALGORITHM H. Apply the standard iterative nonlinear least squares algorithm (even if the f_i are linear), but with metrical Winsorization of the residuals: in each step we replace y_i by

$$f_i(\theta^{(m)}) + \psi(y_i - f_i(\theta^{(m)})).$$

ALGORITHM W. Apply the ordinary weighted least squares algorithm, with weights

$$w_i = \frac{\psi(\Delta_i)}{\Delta_i}$$

determined from the current values of the residuals.

R. Dutter (1975a), (1975b), (1976) has investigated these algorithms and some of their variants.

If very accurate (numerically exact) solutions are wanted, one should use Algorithm S. On the average, it reached the exact values within about 10 iterations in our Monte Carlo experiments. However, these iterations are relatively slow, since elaborate safeguards are needed to prevent potentially catastrophic oscillations in the nonfinal iteration steps, where $\theta^{(m)}$ may still be far from the truth.

Algorithms H and W are much simpler to program, converge as they stand, albeit slowly, but reach a statistically satisfactory accuracy also within about 10 (now faster) iterative steps.

The overall performance of H and W is almost equally good; for linear problems, H may have a slight advantage, since its normal equations matrix C^TC can be calculated once and for all, whereas for W it must be recalculated (or at least updated) at each iteration. By the way, the actual calculation of C^TC can be circumvented just as in the classical least squares case, cf. Lawson and Hanson (1974).

In detail, Algorithm H can be defined as follows.

We need starting values $\theta^{(0)}$, $\sigma^{(0)}$ for the parameters and a tolerance value $\varepsilon > 0$. Now perform the following steps.

1. Put $m = 0$.
2. Compute residuals $\Delta_i^{(m)} = y_i - f_i(\theta^{(m)})$.
3. Compute a new value for σ by

$$(\sigma^{(m+1)})^2 = \frac{1}{a} \sum \chi\left(\frac{\Delta_i^{(m)}}{\sigma^{(m)}}\right)(\sigma^{(m)})^2$$

4. "Winsorize" the residuals:

$$z_i = \psi\left(\frac{\Delta_i^{(m)}}{\sigma^{(m+1)}}\right)\sigma^{(m+1)}, \qquad\qquad i = 1, \cdots, n.$$

5. Compute the partial derivatives

$$c_{ik} = \frac{\partial f_i(\theta^{(m)})}{\partial \theta_k}.$$

6. Solve

$$C^TC\hat{\tau} = C^T\mathbf{z}$$

for $\hat{\tau}$.

7. Put

$$\theta^{(m+1)} = \theta^{(m)} + q\hat{\tau}$$

where $0 < q < 2$ is an arbitrary (fixed) relaxation factor.

8. Stop iterating and go to step 9 if the parameters change by less than ε times their standard deviation, i.e. if for all j

$$|\hat{\tau}_j| < \varepsilon \sqrt{\bar{c}_{jj}}\sigma^{(m+1)}$$

where $\bar{c}_{jj}$ is the jth diagonal element of the matrix $\bar{C} = (C^TC)^{-1}$; otherwise put $m := m + 1$ and go to step 2.

9. Estimate θ by $\theta^{(m+1)}$,

$$\text{var } (y_i) \text{ by } (\sigma^{(m+1)})^2,$$

$$\text{var } (\hat{\theta}) \text{ by } K^2 \frac{1}{n-p} \sum z_i^2 (C^T C)^{-1}.$$

Here K is a correction factor (Huber (1973, p. 812ff)):

$$K = \frac{1 + (p/n) \text{ var } \psi'/(\text{ave } \psi')^2}{\text{ave } \psi'}$$

where

$$\text{ave } \psi' = \frac{1}{n} \sum \psi' \left(\frac{\Delta_i}{\sigma^{(m+1)}} \right) = \mu,$$

$$\text{var } \psi' = \frac{1}{n} \sum \left[\psi' \left(\frac{\Delta_i}{\sigma^{(m+1)}} \right) - \mu \right]^2.$$

In the case of "proposal 2", $\psi(x) = [x]_{-c}^{+c}$, these expressions simplify: μ then is the fraction of non-Winsorized residuals, and

$$K = \frac{1 + (p/n)(1 - \mu)/\mu}{\mu}.$$

Under mild assumptions, this algorithm converges (see Huber and Dutter (1974)): assume that $\rho(x)/x$ is convex for $x < 0$ and concave for $x > 0$, that $0 < \rho'' \leq 1$, and that the f_i are linear. Then one can show that

$$Q(\theta^{(m)}, \sigma^{(m)}) \geq Q(\theta^{(m)}, \sigma^{(m+1)}) \geq Q(\theta^{(m+1)}, \sigma^{(m+1)}),$$

and that these inequalities are strict, unless $\sigma^{(m)} = \sigma^{(m+1)}$ or $\theta^{(m)} = \theta^{(m+1)}$ respectively. The sequence $(\theta^{(m)}, \sigma^{(m)})$ has at least one accumulation point (this follows from a standard compactness argument), and every accumulation point minimizes $Q(\theta, \sigma)$. If the minimum is unique, this of course implies convergence.

Remark. The selection of starting values $(\theta^{(0)}, \sigma^{(0)})$ presents some problem. In the general case with nonlinear f_i's, no blanket rules can be given, except that values of $\sigma^{(0)}$ which are too small should better be avoided. So we might just as well determine $\sigma^{(0)}$ from $\theta^{(0)}$ as

$$(\sigma^{(0)})^2 = \frac{1}{n-p} \sum (\Delta_i^{(0)})^2.$$

In the one-dimensional location case we know that the sample mean is a very poor start if there are wild observations, but that the sample median is such a good one, that one step of iteration suffices for all practical purposes ($\theta^{(1)}$ and $\hat{\theta}$ are asymptotically equivalent), cf. Andrews et al. (1972).

Unfortunately, in the general regression case the analogue to the sample median (the L_1-estimate) is harder to compute than the estimate $\hat{\theta}$ we are interested in, so we shall, despite its poor properties, use the analogue to the sample mean (the L_2 or least squares estimate) to start the iteration. Most customers will want to see the least squares result anyway!

15. Robust covariances: the affinely invariant case. Covariance matrices and the associated ellipsoids are often used to describe the overall shape of points in p-dimensional Euclidean space (principal component and factor analysis, discriminant analysis, etc.). But because of their high outlier sensitivity they are not particularly well suited for this purpose. Let us look first at affinely invariant robust alternatives.

Take a fixed spherically symmetric probability density f in R^p. We apply arbitrary nondegenerate affine transformations $\mathbf{x} \to V(\mathbf{x} - \boldsymbol{\xi})$ and obtain a family of "elliptical" densities

$$(15.1) \qquad f(\mathbf{x}; \boldsymbol{\xi}, V) = |\det V| f(|V(\mathbf{x} - \boldsymbol{\xi})|),$$

let us assume that our data obeys an underlying model of this type; the problem is to estimate the p-vector $\boldsymbol{\xi}$ and the $p \times p$-matrix V from n observations $\mathbf{x}_1, \cdots, \mathbf{x}_n$; $\mathbf{x}_i \in R^p$.

Evidently V is not uniquely identifiable (it can be multiplied by an arbitrary orthogonal matrix from the left), but $V^T V$ is. We can also enforce uniqueness of V by requiring that, e.g., V is positive definite symmetric, or lower triangular with a positive diagonal. Usually, we shall adopt the latter convention. The matrix

$$(V^T V)^{-1},$$

shall be called (pseudo-) covariance matrix; it transforms like an ordinary covariance matrix under affine transformations.

The maximum likelihood estimate of $\boldsymbol{\xi}$, V is obtained by maximizing

$$\log |\det V| + \text{ave} \{\log f(|V(\mathbf{x} - \boldsymbol{\xi})|)\},$$

where ave $\{\cdots\}$ denotes the average taken over the sample.

By taking derivatives, we obtain the following system of equations for $\boldsymbol{\xi}$, V:

$$(15.2) \qquad \text{ave} \{w(|\mathbf{y}|)\mathbf{y}\} = 0,$$

$$(15.3) \qquad \text{ave} \left\{ u(|\mathbf{y}|)\frac{\mathbf{y}\mathbf{y}^T}{|\mathbf{y}|^2} - v(|\mathbf{y}|)I \right\} = 0$$

with

$$\mathbf{y} = V(\mathbf{x} - \boldsymbol{\xi})$$

and

$$w(r) = -\frac{f'(r)}{rf(r)},$$

$$(15.4) \qquad u(r) = -\frac{rf'(r)}{f(r)},$$

$$v(r) = 1.$$

(The reason for introducing the function v shall be explained later.) I is the $p \times p$ identity matrix.

Note that (15.2) can also be written as

(15.2′)
$$\xi = \frac{\text{ave}\,\{w(|\mathbf{y}|)\mathbf{x}\}}{\text{ave}\,\{w(|\mathbf{y}|)\}},$$

i.e. as a weighted mean, with weights depending on the sample, and (15.3) similarly can be reworked into a weighted covariance

(15.3′)
$$(V^T V)^{-1} = \frac{\text{ave}\,\{[u(|\mathbf{y}|)/|\mathbf{y}|^2](\mathbf{x}-\xi)(\mathbf{x}-\xi)^T\}}{\text{ave}\,\{v(|\mathbf{y}|)\}}.$$

(Note that for the multivariate normal density all the weights in (15.3′) are identically 1, so we get the ordinary covariance in this case.)

Equation (15.3), with an arbitrary v, is in a certain sense the most general form for an affinely invariant M-estimate of covariance. I shall briefly sketch why this is so. Forget about location for the moment and assume $\xi = 0$. We need $p(p+1)/2$ equations for the unique components of V, so assume our M-estimate to be determined from

(15.5)
$$\text{ave}\,\{\Psi(V\mathbf{x})\} = 0,$$

where Ψ is an essentially arbitrary function from R^p into the space of symmetric $p \times p$ -matrices.

If all solutions V of (15.5) give the same $(V^T V)^{-1}$, then the latter automatically transforms in the proper way under linear transformations of the $\mathbf{x}$. Moreover, if S is any orthogonal matrix, then one shows easily that

$$\Psi_S(\mathbf{x}) = S^T \Psi(S\mathbf{x})S,$$

when substituted for Ψ in (15.5), determines the same $(V^T V)^{-1}$.

Now average over the orthogonal group:

$$\bar{\Psi}(\mathbf{x}) = \underset{S}{\text{ave}}\,\Psi_S(\mathbf{x});$$

then every solution V of (15.5) also solves

$$\text{ave}\,\{\bar{\Psi}(V\mathbf{x})\} = 0.$$

Note that $\bar{\Psi}$ is invariant under orthogonal transformations

$$\bar{\Psi}_S(\mathbf{x}) = S^T \bar{\Psi}(S\mathbf{x})S = \bar{\Psi}(\mathbf{x}),$$

and this implies that

$$\bar{\Psi}(\mathbf{x}) = u(|\mathbf{x}|)\frac{\mathbf{x}\mathbf{x}^T}{|\mathbf{x}|^2} - v(|\mathbf{x}|)I$$

for some functions u, v. (This last result was found independently also by W. Stahel.)

Influence function. Assume that the true underlying distribution F is spherically symmetric, so that the true values of the estimates are $\xi = 0$, $V = I$. The influence functions for estimators $\hat{\xi}$, $\hat{V}$ computed from equations of the form

(15.2), (15.3), (u, v, w need not be related to f) then are vector and matrix valued of course, but otherwise they can be determined as usual. One obtains

$$IC(\mathbf{x}; F, \hat{\boldsymbol{\xi}}) = \gamma w(|\mathbf{x}|)\mathbf{x},$$

$$IC(\mathbf{x}; F, \hat{V}) = S,$$

with

$$\frac{1}{p}\operatorname{tr} S = -\alpha\left(\frac{1}{p}u(|\mathbf{x}|) - v(|\mathbf{x}|)\right),$$

$$S_{jj} - \frac{1}{p}\operatorname{tr} S = -\frac{p+2}{2}\beta\left(u(|\mathbf{x}|)\left(\frac{x_j^2}{|\mathbf{x}|^2} - \frac{1}{p}\right)\right),$$

$$S_{jk} = -(p+2)\beta u(|\mathbf{x}|)\frac{x_j x_k}{|\mathbf{x}|^2}, \quad j > k,$$

$$= 0, \qquad\qquad j < k.$$

for some constants α, β, γ:

$$\alpha^{-1} = E\left\{\frac{1}{p}u'(|\mathbf{x}|)|\mathbf{x}| - v'(|\mathbf{x}|)|\mathbf{x}|\right\},$$

$$\beta^{-1} = E\left\{\frac{1}{p}u'(|\mathbf{x}|)|\mathbf{x}| + u(|\mathbf{x}|)\right\},$$

$$\gamma^{-1} = E\left\{w(|\mathbf{x}|) + \frac{1}{p}w'(|\mathbf{x}|)|\mathbf{x}|\right\}.$$

For the pseudo-covariance $(V^T V)^{-1}$ the influence function is, of course, $-(S + S^T)$.

The asymptotic variances and covariances of the estimates (normalized by multiplication with $\sqrt{n}$) coincide with those of their influence functions and thus can be calculated easily. For proofs (in a slightly more restricted framework) see Maronna (1976).

Least informative distributions. Let

$$\mathcal{P}_\varepsilon = \{f | f = (1-\varepsilon)\varphi + \varepsilon h, h \text{ spherically symmetric}\}$$

where φ is the standard p-variate normal density.

(i) *Location.* Assume that $\boldsymbol{\xi}$ depends differentiably on a real valued parameter t. Then minimizing Fisher information over $\mathcal{P}_\varepsilon$ with respect to t amounts to minimizing

$$E_F\left[\left(\frac{f'(r)}{f(r)}\right)^2\right].$$

For $p = 1$, this extremal problem was solved in § 12. For $p > 1$, the solutions are much more complicated (they can be expressed in terms of Bessel and Neumann functions); somewhat surprisingly, $-\log f(|\mathbf{x}|)$ is no longer convex (which leads to complications with consistency proofs). In fact, the qualitative behavior of the solutions for large p and ε is not known (it seems that in general there is a fixed sphere around the origin not containing any contamination even if $\varepsilon \to 1$).

For our present purposes however, the exact choice of the location estimate does not really matter, provided it is robust: $w(r)$ should be 1 in some neighborhood of the origin and then decrease so that $w(r)r$ stays bounded.

(ii) Scale. Assume that V depends differentiably on a real valued parameter t. Then minimizing Fisher information over $\mathscr{P}_\varepsilon$ with respect to t amounts to minimizing

$$E_F\left[\left(\frac{rf'(r)}{f(r)}\right)^2\right].$$

In this case, the variational problem has a nice and simple solution (see Fig. 5):

$$u(r) = -\frac{rf_0'(r)}{f_0(r)} = a^2, \quad 0 \le r < a,$$

$$= r^2, \quad a \le r \le b,$$

$$= b^2, \quad r > b,$$

where

$$a^2 = (p - \kappa)^+, \qquad b^2 = p + \kappa$$

for some

$$\kappa = \kappa(p, \varepsilon).$$

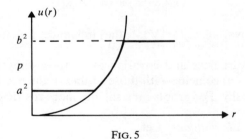

FIG. 5

The corresponding least informative density has a singularity at the origin, if $a > 0$:

$$f_0(r) = (1 - \varepsilon)\varphi(a)\left(\frac{a}{r}\right)^{a^2}, \quad 0 \le r < a,$$

$$= (1 - \varepsilon)\varphi(r), \qquad a \le r \le b,$$

$$= (1 - \varepsilon)\varphi(b)\left(\frac{b}{r}\right)^{b^2}, \quad r > b,$$

where $\varphi(r) = (2\pi)^{-p/2} e^{-r^2/2}$.

Thus, it is rather unlikely that Nature will play its minimax strategy against the statistician, and so the statistician's minimax strategy is too pessimistic (it safeguards against a quite unlikely contingency). Nevertheless, these estimates have some good points in their favor. For instance in dimension 1 the limiting case $\varepsilon \to 1$, $\kappa \to 0$, gives $a = b = 1$ and leads to the median absolute deviation, which is distinguished by its high breakdown point $\varepsilon^* = \frac{1}{2}$.

In higher dimensions, these estimates can be described as follows: adjust a transformation matrix V (and a location parameter ξ, which we shall disregard) until the transformed sample $(\mathbf{y}_1, \cdots, \mathbf{y}_n)$, with $\mathbf{y}_i = V(\mathbf{x}_i - \xi)$, has the following property: if the $\mathbf{y}$-sample is metrically Winsorized by moving all points outside of the spherical shell $a \leq r \leq b$ radially to the nearest surface point of the shell, then the modified $\mathbf{y}$-sample has unit covariance matrix (see Fig. 6).

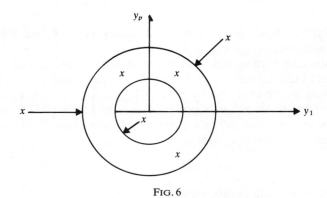

FIG. 6

Breakdown properties. Unfortunately, the situation does not look good in high dimensions p. It appears that $\varepsilon^* \leq 1/p$ for all affinely invariant M-estimates of covariance. This has been proved under the assumption that $u \geq 0$, but in all likelihood is generally true (Huber (1976)).

A note on calculations. Formulas (15.2') and (15.3') can be used to calculate ξ and V iteratively: calculate weights from the current approximations to ξ and V respectively, then use (15.2') and (15.3') to obtain better values for ξ, V. Convergence is not really fast (compare also Maronna (1976, p. 66)), but adequate in low dimensions; a convergence proof is still outstanding. Other computational schemes are under investigation.

16. Robust covariances: the coordinate dependent case. Full affine invariance does not always make sense. Time series problems may be a prime example: one would not want to destroy the natural chronological ordering of the observations. Moreover, the high dimensions involved here would lead to ridiculously low breakdown points.

Several coordinate dependent approaches have been proposed and explored by Gnanadesikan and Kettenring (1972) and by Devlin et al. (1975).

The following is a simple idea which furnishes a unifying treatment of some coordinate dependent approaches.

Let $\mathbf{X} = (X_1, \cdots, X_n)$, $\mathbf{Y} = (Y_1, \cdots, Y_n)$ be two independent vectors, such that $\mathcal{L}(\mathbf{Y})$ is invariant under permutations of the components; nothing is said about the distribution of $\mathbf{X}$. Then the following theorem holds.

THEOREM. *The sample correlation*

$$r(\mathbf{X}, \mathbf{Y}) = \frac{\sum (X_i - \bar{X})(Y_i - \bar{Y})}{[\sum (X_i - \bar{X})^2 \sum (Y_i - \bar{Y})^2]^{1/2}}$$

satisfies

$$Er(\mathbf{X}, \mathbf{Y}) = 0, \qquad E[r(\mathbf{X}, \mathbf{Y})^2] = \frac{1}{n-1}.$$

Proof. Calculate these expectations conditionally, given $\mathbf{X}$, and $\mathbf{Y}$ given up to a random permutation.

Despite this distribution-free result, r obviously is not robust—one single, sufficiently bad outlying pair (X_i, Y_i) can shift r to any value in $(-1, 1)$.

The following is a remedy. Replace $r(\mathbf{x}, \mathbf{y})$ by $r(\mathbf{u}, \mathbf{v})$, where $\mathbf{u}, \mathbf{v}$ are computed from $\mathbf{x}, \mathbf{y}$ according to certain, very general rules. The first two of the following five requirements are essential, the others are merely sometimes convenient.

1. $\mathbf{u}$ is computed from $\mathbf{x}$, $\mathbf{v}$ from $\mathbf{y}$:

$$\mathbf{u} = \Psi(\mathbf{x}), \qquad \mathbf{v} = \Xi(\mathbf{y}).$$

2. Ψ, Ξ commute with permutations.
3. Ψ, Ξ are monotone increasing.
4. $\Psi \equiv \Xi$.
5. $\forall \alpha > 0, \forall \beta, \exists \alpha_1 > 0, \exists \beta_1,$

$$\forall \mathbf{x}, \Psi(\alpha \mathbf{x} + \beta) = \alpha_1 \Psi(\mathbf{x}) + \beta_1.$$

Of these, parts 1 and 2 ensure that $\mathbf{u}, \mathbf{v}$ still satisfy the assumptions of the theorem, if $\mathbf{x}, \mathbf{y}$ did. If part 3 holds, then perfect rank correlations are preserved. Finally, parts 4 and 5 together imply that correlations ± 1 are preserved. In the following two examples, all five assumptions are satisfied.

Example (i). $u_i = a(R_i)$, where R_i is the rank of x_i, and a is a monotone scores function.

Example (ii). $u_i = \psi((x_i - T)/S)$ where ψ is monotone and T, S are any estimates of location and scale.

Some properties of such modified correlations:

1) Let $\mathcal{L}(X, Y) = F = (1 - \varepsilon)G + \varepsilon H$ be a centrosymmetric probability in $\mathbb{R}^2$. Then the correlation satisfies

$$\rho_F(\psi(X), \psi(Y)) \leq (1 - \eta)\rho_G(\psi(X), \psi(Y)) + \eta,$$

$$\rho_F(\psi(X), \psi(Y)) \geq (1 - \eta)\rho_G(\psi(X), \psi(Y)) - \eta.$$

The bounds are sharp, with

$$\eta = \frac{\varepsilon/(1-\varepsilon) \cdot k^2/E_G\psi^2}{1+\varepsilon/(1-\varepsilon) \cdot k^2/E_G\psi^2}$$

where $k = \sup |\psi|$. So η is smallest for $\psi(x) = \text{sign}(x)$, i.e. the quadrant correlation. This should be compared to the minimax bias property of the sample median (§ 11).

2) *Tests for independence.* Take the following test problem. Hypothesis:

$$X = X^* + \delta \cdot Z, \qquad Y = Y^* + \delta \cdot Z_1,$$

where X^*, Y^*, Z, Z_1 are independent symmetric random variables, $\mathscr{L}(X^*) = \mathscr{L}(Y^*)$, $\mathscr{L}(Z) = \mathscr{L}(Z_1)$, δ a real number. The alternative is the same, except that $Z_1 = Z$. Assume that $\mathscr{L}(X^*) = \mathscr{L}(Y^*) \in \mathscr{P}_\varepsilon$ is only approximately known. Asymptotically ($\delta \to 0$, $n \to \infty$) one obtains that the minimum power is maximized using the sample correlation coefficient $r(\psi(\mathbf{x}), \psi(\mathbf{y}))$, where ψ corresponds to the minimax M-estimate of *location*(!) for distributions in $\mathscr{P}_\varepsilon$.

3) *Particular choice for ψ.* Let

$$\psi_c(x) = 2\Phi\left(\frac{x}{c}\right) - 1 \quad \text{for } c > 0,$$

$$\psi_0(x) = \text{sign}(x).$$

where Φ is the standard normal cumulative. If

$$\mathscr{L}(X, Y) = \mathscr{N}\left(0, \begin{pmatrix} 1 & \beta \\ \beta & 1 \end{pmatrix}\right),$$

then

$$E(\psi_c(X)\psi_c(Y)) = \frac{2}{\pi} \arcsin\left(\frac{\beta}{1+c^2}\right).$$

So, for this choice of ψ, at the normal model, we have a particularly simple transformation from the covariance of $\psi(X)$, $\psi(Y)$ to that of X, Y. (But note: if this transformation is applied to a covariance matrix, it may destroy positive definiteness.)

CHAPTER VI

Finite Sample Minimax Theory

17. Robust tests and capacities. Do asymptotic robustness results carry over to small samples? This is not at all evident: 1% contamination means something entirely different when the sample size is 1000 (and there are about 10 outliers per sample) than when it is 5 (and 19 out of 20 samples are outlier-free).

Let us begin with testing, where the situation is still fairly simple. The Neyman–Pearson lemma is clearly nonrobust, since a single sour observation can determine whether

$$\sum_{i=1}^{n} \log \frac{p_1(x_i)}{p_0(x_i)} \gtrless C,$$

if $\log p_1(x)/p_0(x)$ happens to be unbounded.

The simplest possible robustification is achieved by censoring the summands in (17.1):

$$(17.1) \qquad \log \pi(x_i) = \left[\log \frac{p_1(x_i)}{p_0(x_i)} \right]_{c'}^{c''}$$

and basing the decision on whether

$$\sum \log \pi(x_i) \gtrless C.$$

It turns out that this leads to exact, finite sample minimax tests for quite general neighborhoods of P_0, P_1:

$$(17.2) \qquad \mathscr{P}_j = \{P \mid d_*(P_j, P) \leq \varepsilon\},$$

where d_* is either distance in total variation, ε-contamination, Kolmogorov distance, or Lévy distance.

Intuitively speaking, we blow the simple hypotheses P_j up to composite hypotheses $\mathscr{P}_j$, and we are seeking a pair $Q_j \in \mathscr{P}_j$ of closest neighbors, making the testing problem hardest (see Fig. 7).

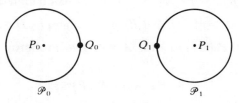

Fig. 7

49

If the likelihood ratio $\pi(x) = q_1(x)/q_0(x)$ between a certain pair $Q_j \in \mathcal{P}_j$ satisfies for all t,

$$(17.3) \qquad\qquad \sup_{P \in \mathcal{P}_0} P\{\pi(x) > t\} = Q_0\{\pi(x) > t\},$$

$$(17.4) \qquad\qquad \inf_{P \in \mathcal{P}_1} P\{\pi(x) > t\} = Q_1\{\pi(x) > t\},$$

then, clearly, the one-sample Neyman–Pearson tests between Q_0 and Q, are minimax tests between $\mathcal{P}_0$ and $\mathcal{P}_1$. One easily proves that this property carries over to any sample size. Note that it is equivalent to: for $P \in \mathcal{P}_0$, $\log \pi$ is stochastically largest when $P = Q_0$; hence $\sum \log \pi(x_i)$, with $\mathcal{L}(x_i) \in \mathcal{P}_0$, becomes stochastically largest when $\mathcal{L}(x_i) = Q_0$ (see, e.g. Lehmann (1959, Lem. 1, p. 73)).

The existence of such a least favorable pair is not self-evident, and it was in fact a great surprise that the "usual" sets $\mathcal{P}_j$ all possessed it, and that the likelihood ratio $\pi(x)$ even had a simple structure (17.1) (Huber (1965)).

This has to do with the following: the "usual" neighborhoods $\mathcal{P}$ can be described in terms of a two-alternating capacity v, that is

$$(17.5) \qquad\qquad \mathcal{P} = \{P \in \mathcal{M} | \forall A, P(A) \leq v(A)\}$$

where v is a set function satisfying (Ω being a complete, separable metrizable space, A, B, being Borel subsets of Ω):

  (i) $v(\phi) = 0, \qquad v(\Omega) = 1,$

  (ii) $A \subset B \Rightarrow v(A) < v(B),$

  (iii) $A_n \uparrow A \Rightarrow v(A_n) \uparrow v(A),$

  (iv) $A_n \downarrow A, A_n$ closed $\Rightarrow v(A_n) \downarrow v(A),$

  (v) $v(A \cup B) + v(A \cap B) \leq v(A) + v(B).$

The crucial property is the last one (the definition of a 2-alternating set function), and it is essentially equivalent to the following: if $A_1 \subset A_2 \subset \cdots \subset A_n$ is any increasing sequence then there is a $P \in \mathcal{P}$ such that for all i, $P(A_i) = v(A_i)$. This simultaneous maximizing over a monotone family of sets occurs in (17.3) and is needed for the minimax property to hold.

Examples. Assume that Ω is a finite set, for simplicity, and let P_0 be a fixed probability.

  (i) $v(A) = (1 - \varepsilon)P_0(A) + \varepsilon$ for $A \neq \phi$ gives ε-contamination neighborhoods:

$$\mathcal{P} = \{P \leq v\} = \{P \in \mathcal{M} | P = (1 - \varepsilon)P_0 + \varepsilon H, H \in \mathcal{M}\}.$$

  (ii) $v(A) = \min(1, P_0(A^\varepsilon) + \varepsilon)$ for $A \neq \phi$ give Prohorov distance neighborhoods:

$$\mathcal{P} = \{P \in \mathcal{M} | d_{Pr}(P_0, P) \leq \varepsilon\}.$$

For further details, see Huber and Strassen (1973).

The theory may have some interesting applications to Bayesian statistics (imprecise priors and imprecise conditional distributions $P(x; \theta)$), compare Huber (1973b)).

18. Finite sample minimax estimation. As a particular case of a finite sample minimax robust test consider testing between $\mathcal{N}(-\mu, 1)$ and $\mathcal{N}(+\mu, 1)$, when these two simple hypotheses are blown up, say, by ε-contamination or Prohorov distance ε. Then the minimax robust test will be based on a test statistic of the form

$$\sum \psi(x_i)$$

with

$$\psi(x) = [x]_{-c}^{+c} = \min(c, \max(-c, x)).$$

This can be used to derive exact finite sample minimax estimates. For instance, assume that you are estimating θ from a sample $x_1, \cdots, x_n$ where the x_i are independent random variables, with

$$\mathcal{L}(x_i - \theta) \in \mathcal{P} = \{F | \sup_t |F(t) - \Phi(t)| \leq \varepsilon\}.$$

We want to find an estimator $T = T(x_1, \cdots, x_n)$ such that for a given $a > 0$ the following quantity is minimized:

$$\sup_{\theta \in \mathbb{R}, P \in \mathcal{P}} \max(P\{T < \theta - a\}, P\{T > \theta + a\}).$$

Then the solution is found by first deriving a minimax robust test between $\theta = -a$ and $\theta = +a$, and then transforming this test into an estimate—find that shift of the original sample for which the test is least able to decide between the two hypotheses. The resulting estimate can be described as follows: let T^*, T^{**} be the smallest and the largest solution of

$$\sum \psi(x_i - t) = 0$$

respectively, with $\psi(x) = [x]_{-c}^{+c}$, $c = c(a, \varepsilon)$, but not depending on n. Then put $T = T^*$ or $T = T^{**}$ at random with equal probability (the more familiar resolution of the dilemma, $T = (T^* + T^{**})/2$ is *not* minimax).

Formally, this is the same kind of estimate as the minimax asymptotic variance M-estimate under symmetric contamination. But note that in the present case we do *not* assume symmetry—on the contrary, the finite sample minimax solution for symmetric distributions is unknown.

For details, see Huber (1968). The question whether this finite sample minimax theory also has an exact, scale invariant counterpart is still open.

CHAPTER VII

Adaptive Estimates

19. Adaptive estimates. Jaeckel (1971b) proposed to estimate location with a trimmed mean $\bar{X}_{\hat{\alpha}}$ whose trimming rate $\hat{\alpha}$ is estimated from the sample itself, namely such that the estimated variance of $\bar{X}_{\hat{\alpha}}$ (see (10.3)) becomes least possible. He showed that for symmetric distributions this estimate is asymptotically equivalent to $\bar{X}_{\alpha^*}$ where $\bar{X}_{\alpha^*}$ is the α-trimmed mean with the smallest asymptotic variance among all trimmed means with fixed trimming rates, provided the underlying distribution F is symmetric, α is restricted to some range $0 < \alpha_0 \leqq \alpha \leqq \alpha_1 < \frac{1}{2}$, and α^* is unique (this last assumption can be removed if F is symmetric). Of course, this estimate is not particularly good for very short and very long tails (where no trimmed mean is really good).

But worse, if the true underlying F is asymmetric, and α^* is not unique, then $\bar{X}_{\hat{\alpha}}$ will not even converge.

Hogg (1967), (1972) avoids the first objection: he selects among a class which includes some estimates appropriate for very short and very longtailed distributions. Since he selects only among a finite set of estimates, one can also bypass the second objection, namely by using the *conditional* asymptotic distribution, given that the particular estimate has been chosen, without straining the good will of the user too much.

These two types of estimators try to choose, among a limited family of estimates, one which is best adapted to the (unknown) underlying distribution. Following an idea of Ch. Stein (1956), one might go even further and construct "fully adaptive" estimates which reach full asymptotic efficiency over a very wide range of distributions (e.g. all distributions with finite Fisher information).

One of the first attempts in this direction was that by Takeuchi (1971). Roughly, his procedure is as follows. Fix the given sample of size n. Fix $k < n$, take random samples S of size k from your given sample, and find that symmetric linear combination of order statistics (in samples of size k) which do best conditionally, given your original sample:

$$\sum_{1}^{k} a_i x_{(i)}^{S}.$$

Then average these estimates over all subsamples S of size k, to obtain your final estimate.

The procedure is difficult to analyze, especially since Takeuchi does not give explicit rules on how k should depend on n. I suspect that it is not robust:

although, for each *fixed F*, the estimate would seem to be asymptotically efficient, I conjecture that for each *fixed n*, the estimate is poor for sufficiently longtailed *F*.

More recently, Beran (1974), Stone (1975) and Sacks (1975) have described fully efficient versions of R-, M- and L-estimates respectively; for some of these estimates one has indeed

$$\mathcal{L}(\sqrt{n}T_n) \to \mathcal{N}\left(O, \frac{1}{I(F)}\right)$$

for every symmetric distribution *F*.

Essentially, all these fully adaptive procedures estimate first a smoothed version $\hat{\psi}$ of $\psi = -f'/f$, and then use a location estimate based on $\hat{\psi}$, e.g. an M-estimate

$$\sum_1^n \hat{\psi}(x_i - T_n) = 0.$$

Possibly, these estimates may be quite poor for asymmetric *F*, because then the variability of $\hat{\psi}$ might contribute a large variance component to the variance of *T*, possibly of the same order of magnitude as the variance already present in the estimate based on the fixed, true ψ. (For symmetric *F*, and provided $\hat{\psi}$ is forced to be skew symmetric, the variance component in question is asymptotically negligible.) This question should be investigated quantitatively.

Also, it is not clear whether these estimates are robust, cf. the remark on Takeuchi's estimate. But see now a most recent paper by R. Beran (1976).

For a comprehensive review of adaptive estimates, see Hogg (1974).

References

D. F. ANDREWS ET AL. (1972), *Robust Estimates of Location: Survey and Advances*, Princeton University Press, Princeton, NJ.

R. BERAN (1974), *Asymptotically efficient adaptive rank estimates in location models*, Ann. Statist., 2, pp. 63–74.

——— (1976), *An efficient and robust adaptive estimator of location*, manuscript.

P. J. BICKEL AND J. L. HODGES (1967), *The asymptotic theory of Galton's test and a related simple estimate of location*, Ann. Math. Statist., 38, pp. 73–89.

J. R. COLLINS (1976), *Robust estimation of a location parameter in the presence of asymmetry*, Ann. Statist., 4, pp. 68–85.

C. DANIEL AND F. S. WOOD (1971), *Fitting Equations to Data*, John Wiley, New York.

H. E. DANIELS (1976), Presented at the Grenoble Statistics Meeting, 1976.

S. J. DEVLIN, R. GNANADESIKAN AND J. R. KETTENRING (1975), *Robust estimation and outlier detection with correlation coefficients*, Biometrika, 62, pp. 531–545.

R. DUTTER (1975a), *Robust regression: Different approaches to numerical solutions and algorithms*, Res. Rep., no. 6, Fachgruppe für Stat., Eidgenössische Technische Hochschule, Zürich.

——— (1975b), *Numerical solution of robust regression problems: Computational aspects, a comparison*, Res. Rep., no. 7, Fachgruppe für Stat., Eidgenössische Technische Hochschule, Zürich.

——— (1976), *LINWDR: Computer linear robust curve fitting program*, Res. Rep., no. 10, Fachgruppe für Stat., Eidgenössische Technische Hochschule, Zürich.

R. GNANADESIKAN AND J. R. KETTENRING (1972), *Robust estimates, residuals and outlier detection with multiresponse data*, Biometrics, 28, pp. 81–124.

F. R. HAMPEL (1968), *Contributions to the theory of robust estimation*, Ph.D. Thesis, Univ. of Calif., Berkeley.

——— (1971), *A general qualitative definition of robustness*, Ann. Math. Statist., 42, pp. 1887–1896.

——— (1973a), *Robust estimation: A condensed partial survey*, Z. Wahrscheinlichkeitstheorie und Verw. Gebiete, 27, pp. 87–104.

——— (1973b), *Some small sample asymptotics*, Proc. Prague Symposium on Asymptotic Statistics, Prague, 1973.

——— (1974a), *Rejection rules and robust estimates of location: An analysis of some Monte Carlo results*, Proc. European Meeting of Statisticians and 7th Prague Conference on Information Theory, Statistical Decision Functions and Random Processes, Prague, 1974.

——— (1974b), *The influence curve and its role in robust estimation*, J. Amer. Statist. Assoc., 69, pp. 383–393.

——— (1976), *On the breakdown point of some rejection rules with mean*, Res. Rep., no. 11, Fachgruppe für Stat., Eidgenössische Technische Hochschule, Zürich.

R. W. HILL (1977), *Robust regression when there are outliers in the carriers*, Ph.D. Thesis, Harvard Univ., Cambridge, MA.

R. V. HOGG (1967), *Some observations on robust estimation*, J. Amer. Statist. Assoc., 62, pp. 1179–1186.

——— (1972), *More light on the kurtosis and related statistics*, Ibid., 67, pp. 422–424.

——— (1974), *Adaptive robust procedures*, Ibid., 69, pp. 909–927.

P. J. HUBER (1964), *Robust estimation of a location parameter*, Ann. Math. Statist., 35, pp. 73–101.

——— (1965), *A robust version of the probability ratio test*, Ibid., 36, pp. 1753–1758.

P. J. HUBER (1967), *The behavior of maximum likelihood estimates under nonstandard conditions*, Proc. Fifth Berkeley Symposium on Mathematical Statistics and Probability, 1, pp. 73–101.

—— (1968), *Robust confidence limits*, Z. Wahrscheinlichkeitstheorie Verw. Gebiete, 10, pp. 269–278.

—— (1972), *Robust statistics: A review*, Ann. Math. Statist., 43, pp. 1041–1067.

—— (1973a), *Robust regression: Asymptotics, conjectures and Monte Carlo*, Ann. Statist., 1, pp. 799–821.

—— (1973b), *The use of Choquet capacities in statistics*, Bull. Internat. Statist. Inst., Proc. 39th session, 45, pp. 181–191.

P. J. HUBER AND V. STRASSEN (1973), *Minimax tests and the Neyman–Pearson lemma for capacities*, Ann. Statist., 1, pp. 251–263; 2, pp. 223–224.

P. J. HUBER AND R. DUTTER (1974), *Numerical solutions of robust regression problems*, COMPSTAT 1974, Proc. in Computational Statistics, G. Bruckmann, ed., Physika Verlag, Vienna.

P. J. HUBER (1977), *Robust Covariances*, Statistical Decision Theory and Related Topics, II, S. S. Gupta and D. S. Moore, eds., Academic Press, New York.

L. A. JAECKEL (1971a), *Robust estimates of location: Symmetry and asymmetric contamination*, Ann. Math. Statist., 42, pp. 1020–1034.

—— (1971b), *Some flexible estimates of location*, Ann. Math. Statist., 42, pp. 1540–1552.

—— (1972), *Estimating regression coefficients by minimizing the dispersion of the residuals*, Ibid., 43, pp. 1449–1458.

J. JUREČKOVÁ (1971), *Nonparametric estimates of regression coefficients*, Ibid., 42, pp. 1328–1338.

L. KANTOROVIČ AND G. RUBINSTEIN (1958), *On a space of completely additive functions*, Vestnik, Leningrad Univ., 13, no. 7 (Ser. Mat. Astr. 2), pp. 52–59, in Russian.

C. L. LAWSON AND R. J. HANSON (1974), *Solving Least Squares Problems*, Prentice-Hall, Englewood Cliffs, NJ.

E. L. LEHMANN (1959), *Testing Statistical Hypotheses*, John Wiley, New York.

R. A. MARONNA (1976), *Robust M-estimators of multivariate location and scatter*, Ann. Statist., 4, pp. 51–67.

R. MILLER (1964), *A trustworthly jackknife*, Ánn. Math. Statist., 35, pp. 1594–1605.

—— (1974), *The jackknife—a review*, Biometrika, 61, pp.

J. A. REEDS (1976), *On the definition of Von Mises functionals*, Ph.D. Thesis, Dept. of Statistics, Harvard Univ., Cambridge, MA.

J. SACKS AND D. YLVISAKER (1972), *Three examples of the nonrobustness of linear functions of order statistics*, manuscript.

J. SACKS (1975), *An asymptotically efficient sequence of estimators of a location parameter*, Ann. Statist., 3, pp. 285–298.

CH. STEIN (1956), *Efficient nonparametric testing and estimation*, Proc. Third Berkeley Symposium on Mathematical Statistics and Probability 1, University of California Press, Berkeley, pp. 187–196.

S. STIGLER (1973), *Simon Newcomb, Percy Daniell and the history of robust estimation 1885–1920*, J. Amer. Statist. Assoc., 68, pp. 872–879.

CH. J. STONE (1975), *Adaptive maximum likelihood estimators of a location parameter*, Ann. Statist., 3, pp. 267–284.

V. STRASSEN (1965), *The existence of probability measures with given marginals*, Ann. Math. Statist., 36, pp. 423–439.

K. TAKEUCHI (1971), *A uniformly asymptotically efficient estimator of a location parameter*, J. Amer. Statist. Assoc., 66, pp. 292–301.

J. W. TUKEY (1960), *A survey of sampling from contaminated distributions*, Contributions to Probability and Statistics, I. Olkin, ed., Stanford University Press, Stanford, CA.

—— (1970), *Exploratory Data Analysis*, Addison-Wesley, Reading, MA.

R. VON MISES (1937), *Sur les fonctions statistiques*, Conf. de la Réunion Internationale des Math., Gauthier-Villars, Paris, pp. 1–8, also in Selecta R. von Mises, Vol. II, pp. 388–394.

—— (1947), *On the asymptotic distribution of differentiable statistical functions*, Ann. Math. Statist., 18, pp. 309–348.